CRITICAL STEPS

CRITICAL STEPS

Managing What Must Go Right
in High-Risk Operations

Tony Muschara
Ron Farris
Jim Marinus

CRC Press
Taylor & Francis Group
Boca Raton London

CRC Press is an imprint of the
Taylor & Francis Group, an **Informa** business

First edition published 2022
by CRC Press
6000 Broken Sound Parkway NW, Suite 300, Boca Raton, FL 33487–2742

and by CRC Press
2 Park Square, Milton Park, Abingdon, Oxon, OX14 4RN

© 2022 Tony Muschara, Ron Farris and Jim Marinus

First edition published by CRC Press 2022

CRC Press is an imprint of Taylor & Francis Group, LLC

Cover design by Joey Thompson.

ISBN: 978-1-032-11429-3 (hbk)
ISBN: 978-1-032-11507-8 (pbk)
ISBN: 978-1-003-22021-3 (ebk)

DOI: 10.1201/9781003220213

Typeset in Times
by Apex CoVantage, LLC

Contents

Acknowledgments

Since starting this project more than 2 years ago, we have been greatly helped in the preparation and review of this book, which we think will save lives and preserve livelihoods. We wish to express our appreciation to the following friends, colleagues, and associates for helping steer the content of this book in the right direction: Juan Alvarez, Bill Blunt, Earl Carnes, Rosa Carrillo, Mike Donovan, John Ensor, Dave Fink, John Hentges, Luke Kelvington, Lisa Lande, William Mairson, Jack Martin, James Merlo, Tom Neary, Steve Newell, David Provan, Lee Reynolds, Rizwan Shah, Steve Sondergard, Thomas Swanson, Amy Wilson, John Wreathall, David Wulf, and Marilyn Yvon (editor).

Additionally, we want to express our innermost appreciation for the patience of our spouses who were with us along the way, giving us space and time to immerse ourselves in this project. With love and gratitude, we dedicate this book to **Pam Muschara**, **Lisa Gardea-Farris**, and **Debbie Merica**.

Authors

Tony Muschara, CPT, is a specialist in the field of human and organizational performance (H&OP), and principal consultant and owner of Muschara Error Management Consulting, LLC since 2008. Tony is a Certified Performance Technologist (CPT) awarded by the International Society for Performance Improvement (2006). Tony is the author of *Risk-Based Thinking: Managing the Uncertainty of Human Error in Operations* (2018). He authored several nuclear industry publications while employed with the Institute of Nuclear Power Operations (INPO), most notably *Excellence in Human Performance* (1997) and the *Human Performance Reference Manual* (2006), the latter adopted with revisions by the U.S. Department of Energy as the *Human Performance Improvement Handbook* (Volume 1). Tony served 7 years on active duty in the U.S. Navy submarine service (1975–1982), and while on active duty, Tony qualified in submarines (1978) and certified as an Engineer of Naval Nuclear Propulsion Plants by Naval Reactors (1979). Tony earned a Master of Business Administration (MBA) degree from Kennesaw State University (1989) near Atlanta, Georgia, and a Bachelor of Science degree in general engineering (mechanical) from the United States Naval Academy (1975) in Annapolis, Maryland.

Ron Farris is a H&OP specialist, principal consultant and business partner of HOPE Consulting, LLC since 2016, and the Chief Operations Officer at High Reliability Training, Inc. He has authored several department human factors publications and technical documents while employed with the Idaho National Laboratory (INL). Ron was previously an adjunct professor at the University of Idaho teaching courses in support of Industrial Technology degree programs and Human Performance Certificate program. He has provided both practical and classroom support for the Nuclear Regulatory Commission (NRC), Electrical Power Research Institute (EPRI), Department of Energy (DOE), commercial nuclear and fossil fuel energy sectors, mining, and the fuel and petrochemical industry. Ron spent 27 years at the INL where he was an accident investigator, safety engineer, manager of the Center for Human Performance Improvement, and human factors research scientist. He was a senior reactor operator at Argonne National Laboratory's Experimental Breeder Reactor II. Ron served 8 years on active duty in the U.S. Navy nuclear program (1981–1989). While on active duty, he qualified on four different Navy reactor types and as a Chief Machinery Operator (1985) while aboard the USS Nimitz aircraft carrier. Ron earned a Master of Science degree (MS) with a focus on industrial safety from the University of Idaho (2006) Moscow, Idaho.

Jim Marinus specializes in high-risk operations management, high reliability, and resilience, and is principal consultant and owner of Jamar Operations, LLC (2015–present). When not consulting, Jim is actively involved with the international communities of practice for H&OP, high reliability, and resilience. Prior to his time as a management consultant, his professional career spanned time with the U.S.

Department of Energy (DOE) (1983–2012) and in the U.S. Navy nuclear submarine service (1974–1983). While employed with the DOE, Jim directed high-risk
and multidisciplinary research operations at the Idaho National Laboratory. He also
assisted the National Laboratory community with application of the practices and
principles of high reliability, H&OP, and safety culture. Early on, Jim assisted leaders at DOE's Washington, D.C. headquarters with the development and implementation of technical requirements and guidelines that spanned the operations, weapons,
global security, science/technology, ESH, radiation protection, and training communities. On temporary assignment to the Institute of Nuclear Power Operations
in Atlanta, Georgia (1986–1987), he assisted with the training and accreditation of
power plant training programs. Way back, Jim rode nuclear-powered submarines as
a mechanical equipment operator. He earned a Bachelor of Science degree in nuclear
technology from Excelsior College (1996) in Albany, New York and is a Registered
Radiation Protection Technologist (1984).

List of Figures

List of Tables

Introduction

What activities, if performed less than adequately, pose the greatest risks to the well-being of the system?[1]

—Dr. James Reason
Author and Professor Emeritus

Unlike people in most other organizations, however, HROs* have a good sense of what needs to go right and a clearer understanding of the factors that signal that things are unraveling.[2]

—Drs. Karl Weick and Kathleen Sutcliffe
Authors: *Managing the Unexpected*

Absolute safety is impossible to achieve, especially with any human endeavor.[3] Absolute safety exists only when a system, device, product, or material can *never* cause or have the potential to cause harm. Risk arises otherwise. Safety is thought to exist when there is an "acceptable risk" for a particular operation, where work occurs consistently and predictably without harm to assets—person, property, product, or environment, among others. Many of society's operations are operator-dependent, which suggests that these operations are inherently risky—people are fallible; they make mistakes. As the following tragedy reveals, safety and survival are often in the hands of people on the front lines doing normal work every day.

DEADLY MEDICATION ERROR[4]

The Nurse providing initial care of a troubled, pregnant 16-year-old mistakenly injected an epidural painkiller instead of antibiotic directly into her bloodstream. The young girl was ready to give birth to her baby. Her heart stopped beating and the girl could not be revived. However, the baby was delivered successfully by emergency caesarean section. How could a registered nurse with years of experience in this women's unit do such a thing? Easily.

The veteran Nurse bypassed several safety practices. However, most organizational systems required to support the practices were flawed, encouraging nurses to work around them to deliver care. These system flaws conspired, so to speak, contributing to her fatal mistake. Here are the facts.

- The Nurse had 15 years of experience working in the obstetric unit with an unblemished work record.
- At the time of the incident, the Nurse had worked two consecutive eight-hour shifts the day before and slept in the hospital before coming

---

* High-Reliability Organizations.

on duty the following morning. Hospital management encouraged nurses to work long hours, rewarding those with the most overtime hours each year with a free professional development trip.

- The Nurse was working with two patients concurrently, both involving emotional trauma. The other patient was in labor about to deliver a deceased baby. The Nurse's supervisor assigned this patient to her because she ran the hospital's grief program.
- The 16-year-old was suffering from a strep infection and was afraid and crying. She had no prenatal care. One report says she was terrified.
- A doctor prescribed an antibiotic to protect the unborn baby from the mother's strep infection.
- A patient identification wrist band had been prepared and placed in the pocket of the teen's medical chart. However, the Nurse did not place the bracelet on the girl promptly as her focus was on alleviating the teen's fears and anxiety.
- The Nurse did not use the hospital's new bar-coding system for intravenous (IV) fluids, installed a couple of weeks before, designed to match the right medication to the right patient. The hospital's nurses, many of whom had not been trained on the system, often bypassed the system because software glitches hampered its reliability. It failed to register IV bags 30–70 percent of the time.
- Knowing an epidural painkiller would be used later during delivery and to ease the teen's anxiety, the Nurse acquired a 100 ml IV bag from the automated medication-dispensing machine to show it to the 16-year-old patient to relieve her fears. Afterward, the Nurse placed the bag on the bedside table, continuing to comfort the young mother-to-be.
- Moments later another nurse delivered a 100 ml IV bag of antibiotic, placing it on the same bedside table alongside the other 100 ml bag. A bag of epidural painkiller looks the same as a bag of antibiotic, with the only differences being the label and an orange dot instead of a yellow dot.
- Both medications were brought into the patient's room before doctors' orders were given, contrary to hospital policy.

While talking to the patient, the Nurse inadvertently picked up and hung the incorrect IV bag on the pole near the girl's bed, thinking it was an antibiotic. Apparently, she inserted the IV tube into the girl's IV access line on her arm, and then opened the tubing clamp, feeding the epidural painkiller directly into the teen's bloodstream—the point of no return.* Later, she was charged by

---

* The exact error is vague. No report or reference accurately describes the physical act the Nurse explicitly did to initiate the flow of epidural painkiller into the vein of the young girl. Common words used to describe her mistake include "gave," "infused," and "injected," all ambiguous. We do know that the critical act is embedded in those words somewhere.

the state with a felony alleging criminal negligence and was subsequently dismissed from the hospital. Charges were later reduced to two misdemeanors. Regardless, her nursing license was suspended. She was barred for several years from working for any hospital. For what it's worth, punishing those who err does not improve safety.

It is impossible from a human perspective to be fully alert and vigilant 100 percent of the time, completely informed, always rational, and error-free—especially for long periods. If you need 100 percent reliability, then you had better use a machine.* However, to date, machines do not have sufficiently sophisticated adaptive capacity (intelligence) to take the place of human operators. Most operations require some degree of human oversight and control of work processes. That means that some human actions MUST go right the first time, every time, where a loss of control is unacceptable. As Gene Kranz, former NASA flight director, is so famously known to have said during the Apollo 13 incident, "Failure is not an option."[5] CRITICAL STEPS address those aspects of work that are intolerant of human error—human single-point failures. *A CRITICAL STEP is any human action that triggers immediate, irreversible, and intolerable harm to something important if that action or a preceding action is performed improperly.*

By design, CRITICAL STEPS happen every day at work and at home. The aim of this book is to help you manage those operational aspects that depend on the human operator getting things right. As tragically illustrated previously, the healthcare system depends greatly on people getting things right the first time every time. Presumably, the Nurse's CRITICAL STEP was opening the tubing clamp on the IV tube that allowed the flow of the epidural painkiller into the bloodstream of the young girl. Nurses open IV clamps every day in hospitals and other healthcare facilities.

Other human actions important to the success and resilience[†] of the performance of CRITICAL STEPS include what are known as Risk-Important Actions (RIAs), the "preceding action" mentioned earlier. One RIA done earlier by the Nurse was inserting the IV tube into the IV access line attached to the girl's arm. That act created the pathway between the IV bag and her bloodstream, protected only by a closed tube clamp. RIAs are not something to avoid. They are necessary to create the conditions for work to occur. Chapter 4 explores their relationship with CRITICAL STEPS and their importance to safety.

---

* We attribute this idea to Dr. Todd Conklin, a prolific H&OP speaker and author.
† Resilience is the intrinsic ability of a system (an organization) to adjust its functioning before, during, and after a challenge, disturbance, or failure to sustain and improve operations under both expected and unexpected conditions. Resilience is the capacity to sustain safety, productivity, quality, etc.—to adapt—in the face of unexpected conditions and challenges, the ability to succeed under varying conditions.

Together, human fallibility and complex systems* cast a shadow of uncertainty over all hazardous operations. The combinations of fallible human beings, designs and procedures based on faulty assumptions, many difficult tasks, complex technologies, the marshalling of hazards, and numerous regulatory prescriptions can make every operation seem critical. Despite our best efforts, events still occur. Organizations and their systems are not always aligned for safety—things change, hidden pitfalls arise, equipment wears out, priorities shift, people make trade-offs when goal conflicts arise, resources are limited, and so on. That's why it's so important to single out the most important human actions that pose the greatest risk, and make sure they go right. But if they don't go right, minimize the harm to the most important assets, to *fail safely*. Therefore, frontline workers need latitude for safety, what is known as *adaptive capacity*, in the workplace, able to respond with some degrees of freedom to unforeseen work situations—to *do* safety and achieve success.

THE PRINCIPAL GOAL OF MANAGING Critical Steps

The principal goal of managing CRITICAL STEPS is to *maximize the success of people's performance in the workplace, creating value without losing control of the built-in hazards necessary to do work.* Dr. Ron Westrum suggested three practical meanings (applications) of resilience, two of which we address in this book, 1) preventing something bad from happening, and 2) preventing something bad from getting worse.[6] The third aspect of resilience, recovering from something bad that happened, is beyond the scope of this book. This book purposefully limits the scope to those frontline aspects of high-risk operations most related to managing the human performance risk, enhancing success during operations: exercising positive control of CRITICAL STEPS and failing safely after losing control.

In practical terms, managing CRITICAL STEPS is a necessary form of operational hazard control, where an alternate means—other than human—of reduction or control of a built-in hazard by design is not available.[7] Occasionally, operations are established for highly innovative systems that have never existed before (such as space exploration), for which there is limited experience. It is near impossible to prescribe detailed design safety solutions for such endeavors.[8] But we still want to make sure the right things go right the first time, every time. The right things are those high-risk actions or processes that create value, that is, work. The concept of CRITICAL STEPS should promote successful operational hazard control even without a procedure. Most errors have no consequences; and, because of their trivial nature, most errors occur without our knowledge. However, what if error *is* unacceptable? What if people's lives and livelihoods are at stake? What if failure really is not an option, as was the case with the Nurse in the medication event? But, what if failure

---

* Complex systems are characterized by the presence of many components, with concurrent and possibly obscure interactions, and individually adaptive components (mostly people and sophisticated software), the effects of which are not easily comprehensible by any one person.

does happen? How should people respond to minimize the harm if they lose control? Later, we discuss preparations to "fail safely" in response to losses of control at CRITICAL STEPS, if practicable.

The practice of managing CRITICAL STEPS changes the emphasis from simple error avoidance to ensuring success proactively and systemically. An added benefit of this practice optimizes efficiencies and productivity as well as safety of operations that are indeed high risk. It provides a structure for everyone in the organization to have ongoing, robust, technical, interdisciplinary conversations about what must go right in the pursuit of business goals. While the practice of managing CRITICAL STEPS focuses on what's most important, it accordingly identifies low-risk work activities where the loss of control has little or no impact. There is no need to prevent all human error. One, it's impossible; two, it's expensive; and three, it distracts from your focus. To us it makes better sense to target high-risk activities, ensuring that they go right, and to avoid the truly high-risk errors. This has the effect of giving control of low-risk operations to the expert judgment of the workforce.

Consider downhill ski racing. Racers have practiced racing techniques for thousands of hours, honing their skills and creating muscle memory. In preparation, they slowly pre-ski the racecourse to evaluate slope, snow and light conditions, and optimal lines. They look for bad snow, trees and other hard objects close to the course,

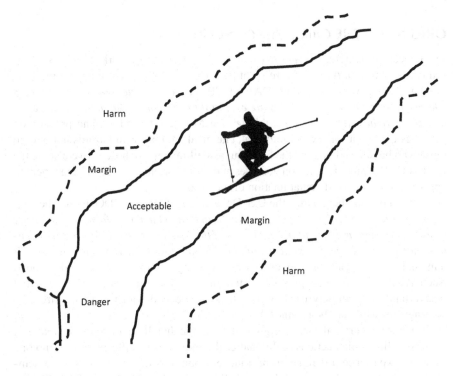

FIGURE I.1   Downhill ski racer.

tricky transitions, and other conditions that could spell failure, making mental notes of when extra caution is prudent. Then they ski hard and fast, following the lines they planned. They look where they want to go, not where they don't. It's a choice, honed by experience. They know that if they look where they don't want to go, it is more likely that they will go there—a phenomenon called "target fixation."[9] They know they must focus on success, the best line, while they remain mindful of incursions into their desirable lane caused by obstacles, and occasionally by chaos due to variabilities in the environment. If you turn your gaze toward an unanticipated danger, you can't see the line of success. Behavior choices in the workplace are a form of CRITICAL STEP management. We will show you how this works, keeping your eye on the goal, exercising positive control, while factoring in information just outside your target view—a persistent sense of unease—developing and taking advantage of expert intuition.

Drs. Karl Weick and Kathleen Sutcliffe state in their third edition of *Managing the Unexpected*, on high-reliability organizations (HROs), "HROs develop capabilities to detect, *contain*, and *bounce back* from those inevitable errors that are part of an uncertain world. The hallmark of an HRO is not that it is error-free but that errors don't disable it"—to fail safely—to keep something bad that has happened from getting worse.[10] In our world of high-hazard, complex, and interconnected operations, it is important to understand how we can organize to enable positive control of CRITICAL STEPS and to minimize harm to assets should control be lost.

ORIGINS OF THE CRITICAL STEP CONCEPT

The concept of CRITICAL STEPS exists in several domains of work. In food service, *critical control points* (CCPs) are identified and controlled. According to the U.S. Food and Drug Administration (FDA), a CCP is *a point in the food service process where controls can be applied and are essential to prevent or eliminate a food safety hazard or reduce it to an acceptable level.*[11] Most food handling and preparation activities contain a variety of biological, chemical, and physical hazards not only to those who handle and prepare the food, but also ultimately to those who consume the food. A CCP is more focused on defenses and thus is oriented in identifying where to apply a control to avoid contamination of the food.

To point out another source, the U.S. Department of Defense (DOD) uses the concept of a *safety-critical function*. This is a *"function, which if performed incorrectly or not performed, may result in death, loss of the system, severe injury, severe occupational illness, or major system damage."*[12] Safety-critical functions include all human, hardware, and software processes necessary to either work when demanded (such as actuating a fire suppression system in case of a fire) or not work if activated inadvertently or untimely (such as interlocks associated with raising an aircraft's landing gear while on the ground).[13]

The phrase "critical step" originated with the handling of nuclear weapons. Understanding which actions really matter the most was significant because no one wants to experience a detonation of a nuclear bomb. Nuclear weapons are assembled, dismantled, and maintained at the U.S. Department of Energy's (DOE) Pantex

Plant near Amarillo, Texas, where "much ado" is paid to CRITICAL STEPS. The facility originally crafted the following definitions[14]:

- *Hazardous step*—a procedure step that, if performed incorrectly, has a "potential" to "immediately" result in a dominant high-energy detonation.
- *Critical step*—a procedure step, that if skipped or performed incorrectly, will increase the "likelihood" of a high-energy detonation . . . at some later step in the procedure.

By these definitions, "critical step" as used by Pantex was less serious than "hazardous step," which seems contradictory. To us, the term CRITICAL STEP has a greater sense of dread—fear and apprehension of an unwanted outcome—than the word hazardous (dangerous or risky). Also, the description of a critical step is similar to what we refer to as an RIA. The real critical step is what the description refers to as "some later step in the procedure."

As you can see, the FDA, DOD, and DOE definitions have similar purposes but leave room for rationalization. For sure, there has never been an unintended detonation of a nuclear weapon; however, we believe the words *likelihood, may, acceptable,* and *potential* lack sufficient specificity and clarity for frontline workers. Since the mid-2000s, Pantex, along with other DOE facilities, incorporated the concept of CRITICAL STEPS in its operations, tailoring it to the nuances of the various technologies and missions of the department and its contractors. Human beings are notoriously inept at estimating probabilities. We usually underestimate the likelihood of an occurrence. That's why we describe CRITICAL STEPS in more concrete terms that minimize doubt as to what can happen. Vague terms, such as mentioned previously, tempt people to rationalize their sense of control—especially in the throes of production pressures. You and your organization are better off adopting more concrete terms that keep the red lights flashing in the minds of those who perform CRITICAL STEPS. Frontline workers must know: 1) what absolutely must go right, anticipating what can go wrong; 2) what to pay attention to (high-risk actions and the asset's *safety-critical parameters*); and 3) what to do before they start their work to respond properly to achieve success and protect assets from harm. The foregoing mental aspects of high-risk work are closely aligned with the cornerstone elements of RISK-BASED THINKING (see description in Appendix 2).

WHO BENEFITS USING CRITICAL STEPS?

The short answer is that anyone involved with hands-on work, in direct contact with key assets and their built-in hazards, benefits from using CRITICAL STEPS. We live in a world where our actions control or moderate the transfer of energy, movement of matter (solids, liquids, or gases), or the transmission of information that can and do cause something of value to happen or to cause harm. Whether we are turning into traffic from a side road on to a busy city street, walking with an infant in our arms down a flight of stairs, making an incision during surgery, starting a high-pressure pump, or clicking "Enter" for financial transactions, each of us performs CRITICAL STEPS every day of our lives, both at home and at work.

CRITICAL STEPS occur naturally in many aspects of high-risk operations, maintenance, engineering, research, and administrative work. People doing physical work perform CRITICAL STEPS as an essential part of their jobs—it's normal and necessary to be successful in the marketplace. However, there is a broad range of functions performed by support staff, usually removed in time and space from the front line, who establish conditions that influence workers' behavior choices in the workplace. This is an organizational aspect of human and organizational performance (H&OP).

Support staff—engineers, scientists, accountants, instructors, procedure writers, and other knowledge workers—who function in the information domains are expected to perform their activities without error. Error is never acceptable, but it is managed by means of redundant checks and reviews. Every engineer or scientist intends to turn out a work product that is 100 percent accurate, that is 100 percent complete, that meets 100 percent of the requirements, and that results in a defect-free product. Regardless, errors by knowledge workers do not trigger immediate harm. One could define a human action that would *eventually* produce harm if that action or subsequent checks downstream the production process were performed improperly. This is a classic latent error, not an active error. Latent errors do not satisfy the CRITICAL STEP definition.

However, there are occasions when knowledge workers perform hands-on work. It is during these times when CRITICAL STEPS are applicable to what they do, whether they are working with physical mockups, running high-energy particle experiments, or executing buy or sell stock trades on an exchange. Sometimes knowledge workers trigger the worst outcomes for organizations.

ORGANIZATION OF THE BOOK

1. Chapter 1 defines CRITICAL STEPS and describes its attributes. This chapter discusses how identifying and controlling CRITICAL STEPS promote organizational success while eliminating harm to the company and its key assets.
2. Chapter 2 orients the reader as to how to think about human performance (frequently abbreviated **Hu*** herein) risk in frontline operations. Most importantly, we reframe human error as a loss of control because events— harm to assets—are caused by hazards, not by frontline workers.
3. Chapter 3 examines the *Work Execution Process*, which describes how work is planned, executed, and improved. Understanding the three phases of work helps in the systematic preparation and performance of CRITICAL STEPS, as well as learning from surprises.
4. Chapter 4 examines those human actions—*Risk-Important Actions*— that precede CRITICAL STEPS that establish the required preconditions,

---

* The abbreviation **Hu** (letters pronounced separately: "aitch u") was adopted by the commercial nuclear energy industry in the mid-1990s as a cause category of events. "HP" had already been adopted to represent "health physics," a domain of knowledge associated with the study of the effects of radiation on human health. Consequently, Hu was used instead of HP. **Hu** is bolded to represent the abbreviated form of human performance of individual performers.

Risk-Important Conditions, to ensure success when frontline workers perform CRITICAL STEPS.

5. Chapter 5 addresses the performance of CRITICAL STEPS in the workplace, emphasizing the need for positive control, shifting between fast and slow thinking, applying RISK-BASED THINKING using **Hu** Tools, providing time-tested methods to manage the leadup to and execution of CRITICAL STEPS.

6. Chapter 6, the longest chapter, guides the reader with a workable strategy for managing CRITICAL STEPS—learning to more effectively and consistently identify and control CRITICAL STEPS from a systems perspective; it includes ways and means for augmenting adaptive capacity to respond to the unexpected.

7. Chapter 7 details the CRITICAL STEP MAPPING (CSM) process to systematically identify perpetual CRITICAL STEPS in existing technical procedures and pinpoint means of their control.

8. Chapter 8 examines the integration and implementation of the principles and practices of managing CRITICAL STEPS into operations while reinforcing systems thinking.

Because a new vocabulary is associated with H&OP and with CRITICAL STEPS in particular, Appendix 1 provides a glossary of terms and phrases commonly encountered in the following pages. You are encouraged to flag this appendix, as you will likely refer to it often during your reading. We refer to H&OP and RISK-BASED THINKING frequently throughout the book. Appendix 2 provides a brief primer on these topics including the principles for managing H&OP. We recommend reading this appendix before starting Chapter 1.

Stories, experiences, and events that reveal the reality of CRITICAL STEPS in ordinary work and in everyday life are spread throughout the book. Most are true or else inspired by actual events, some tragic, some humorous—all denoted with single-line borders. Each chapter begins with an account of an event relevant to that chapter's content. Each chapter concludes with *Key Takeaways* that summarize the most important principles or ideas of the chapter, as well as *Checks for Understanding* with a few quiz questions associated with the chapter's content to confirm your understanding and application of the content. Appendix 3 provides answers to the quiz questions. As with Appendix 1, we suggest flagging that page also for quick access. Finally, with application in mind, each chapter concludes with *Things You Can Do Tomorrow*, practical suggestions on how to apply the concepts described in the respective chapter, with little or no resource allocation.

We hope this book will help you think more systematically about how success happens each day not only in your organizations, but also in your personal and professional lives, ensuring that the CRITICAL STEPS in your life and work result only in adding value.

REFERENCES

1 Reason, J. (1997). *Managing the Risks of Organizational Accidents*. Burlington: Ashgate (p. 91).
2 Weick, K., and Sutcliffe, K. (2001). *Managing the Unexpected*. San Francisco: Jossey-Bass (p. 89).
3 Sgobba, T. (ed. in chief), et al. (2018). *Space Safety and Human Performance*. Cambridge: Butterworth-Heinemann (p. 282).

4 Landro, L. (2010, March 16). 'New Focus on Averting Errors: Hospital Culture.' *Wall Street Journal*. Retrieved from: http://online.wsj.com/article/SB10001424052748704588404575123500096433436.html. And: Institute for Healthcare Improvement, Patient Safety Executive Development Program. Retrieved from: http://app.ihi.org/Events/Attachments/Event-2926/Document-6137/Handout_Julie_Thao_Case_Study.pdf.

5 Kranz, G. (2000). *Failure is Not an Option*. New York: Simon and Schuster (p. 59).

6 Westrum, R. (2006). 'A Typology of Resilience Situations.' In: Hollnagel, E., Woods, D., and Leveson, N. (eds.). *Resilience Engineering: Concepts and Precepts*. Aldershot: Ashgate (p. 59).

7 Sgobba, T. (ed. in chief), et al. (2018). *Space Safety and Human Performance*. Cambridge: Butterworth-Heinemann (p. 290).

8 Ibid. (p. 293).

9 Cummins, D. (2013, September 12). 'Do You Have "Eyes on the Prize" or "Target Fixation?" One Leads to Success, the Other to Disaster.' *Psychology Today*.

10 Weick, K., and Sutcliffe, K. (2015). *Managing the Unexpected* (3rd ed.). Hoboken: Wiley (p. 12).

11 U.S. Food & Drug Administration (updated, 2017, December 19). *Hazard Analysis and Critical Control Point Principles and Application Guidelines*. Retrieved from: www.fda.gov/food/hazard-analysis-critical-control-point-haccp/haccp-principles-application-guidelines.

12 United States Department of Defense (2000). *Standard Practice for System Safety* (MIL-STD 882D). Washington, DC: Government Printing Office.

13 Sgobba, T. (ed. in chief), et al. (2018). *Space Safety and Human Performance*. Cambridge: Butterworth-Heinemann (p. 281).

14 Fischer, S., et al. (1998). *Identification of Process Controls for Nuclear Explosive Operations*, and U.S. Department of Energy (DOE). "Nuclear Explosive Safety Order 452.2A.

1 What Is a CRITICAL STEP?

If an operation has the capacity to do work, then it has the capacity to do harm.*

—**Dorian Conger**

Roger, I was afraid of that. I was really afraid of that.

—**Battalion Commander of a U.S. Army Apache helicopter
flying in the gunner's seat after a "friendly fire" tragedy**

FATAL FRIENDLY FIRE[1]

On February 17, 1991, at approximately 1:00 a.m., a U.S. Bradley Fighting Vehicle and an armored personnel carrier were destroyed by two missiles fired from a U.S. Apache helicopter. Two U.S. soldiers were killed, and six others were wounded. This friendly fire tragedy took place in the Persian Gulf during Operation Desert Storm. The incident occurred after U.S. ground forces, which were deployed along an east-west line about 3 miles north of the Saudi-Iraqi border, reported several enemy sightings north of their positions. In response, ground commanders called for Apache reconnaissance of the area.

Apache cockpits have two sections: the front seat is reserved for the gunner and the back seat for the pilot. The pilot controls the flight pattern, and the gunner engages the target with the helicopter's weapon systems. Both sections of the cockpit have flight and weapons control if one must take control of the other.

Every night for the first couple of weeks of February, battalion helicopters responded to reports from U.S. ground forces of apparent movements of Iraqi vehicles, all false alarms. A U.S. Army Lieutenant Colonel was the Battalion Commander of a U.S. Army Apache helicopter strike force. Just days before, helicopters from the Colonel's battalion misidentified and fired on a U.S. Army scout vehicle, missing it without damage or injury—a near hit.

U.S. armored forces on the ground operating in the area reported possible enemy sightings—suspected Iraqi armored vehicles moving toward a U.S. tank squadron. Commanders of the ground forces asked for aid from the Apache battalion based about 60 miles south of the border to explore the area and to engage them if enemy presence was found. The Colonel with his copilot and two other Apache helicopters responded quickly, urgently directed to patrol an area north of the line of U.S. tanks. Because of an imminent sandstorm with intense winds and low visibility, the Colonel decided to command the lead

---

* This statement is attributed to Dorian Conger, who made this statement to students during the introduction to a MORT cause analysis class. (MORT: Management Oversight Risk Tree).

Apache himself, in the gunner's seat, even though he had only 3 hours of sleep in the previous 24 hours. They launched at 12:22 a.m. Due to the urgency of the request, a normal, detailed premission briefing was not done.

Upon arriving on station at 12:50 a.m., the helicopter's target acquisition system detected the vehicles. Two suspicious vehicles appeared near the eastern end of the line of U.S. ground forces, noting the targets' locations by measuring their distance from the aircraft with a laser beam, automatically entered into the weapons fire control computer. The Colonel estimated the suspicious vehicles were about a quarter mile in front, the first mistake. He misread the grid coordinates of the alleged targets on the helicopter navigation system, reading instead the search coordinates that he manually entered into the navigation system while in route early in the flight. As a result, he misidentified the target vehicles' location as being north of the line of friendly vehicles, which coincidently were in the exact location of previously reported enemy sightings.

A discussion ensued between the three Apache pilots and the ground commander to authenticate their identity. Apache helicopters were not equipped with IFF—an automated system referred to as "Identification Friend or Foe." In the darkness, the vehicles could not otherwise be identified.

The ground commander insisted that no U.S. forces were ahead of the line, that the vehicles must be enemy, and twice replied to the Colonel, "Those are enemy. Go ahead and take them out." Pilots of the other two Apaches also thought the vehicles were enemy. More ominously, there were persistent search-radar alerts being received in the cockpit, adding to the stress of the moment. These alerts, responding to radar emitted by friendly forces, were misidentified by the Apache computers as an enemy system. Even the Colonel's copilot prompted him, "Do em!" more than once. Yet he felt uneasy as to the identity of the vehicles. The Colonel is recorded to have said, "Boy, I'm going to tell you, it's hard to pull this trigger," asking for help to verify current helicopter heading and bearing to and grid coordinates of targets. He states the targets' grid coordinates aloud, again misreading them, the second mistake. No one recognizes the error. His copilot states, "Ready in the back."

The Colonel decided to fire on the vehicles with the Apache's 30-millimeter cannons (machine guns), which would have inflicted less damage than a missile just in case they were friendlies. The gun emitted only a few rounds before jamming (sand). He then fired two Hellfire missiles* at the suspected vehicles—the third, but deadly, mistake. Shortly thereafter, the Apaches received a cease fire order. The missiles had already been fired and both vehicles, a Bradley Fighting Vehicle and an armored personnel carrier, were destroyed, killing two U.S. soldiers inside. The Colonel softly said, "I was afraid of that, I was really afraid of that."

---

* The laser-guided Hellfire missile is the main armament on the Apache helicopter, designed for the destruction of armor and other hardened targets.

The Colonel knew the point of no return: pulling the trigger! He said it. But human fallibility entered the decision-making process, hampered by sleep deprivation, a fierce desert dust storm, inadequate human factors in the cockpit, inferior teamwork, and the stress of combat that worked against him, his team, and even ground commanders. *He did his best* under the circumstances. Would you have done anything different? Be honest. The system and the battlefield worked against him. Sometimes doing your best isn't good enough.

WORK = RISK

When you do work, something changes.* Physical work is the application of force over a distance ($W = f \cdot d$). Work is necessary to create value. Except where automation is used, work requires people to touch things—to oversee, manipulate, record, or alter things. Jobs and tasks comprise a series of human actions designed to change the state of material or information to create outputs—assets that have value in the marketplace. The risk of harm to those assets emerges when people do work, without which nothing of value is created. *Work is energy directed by human beings to create value.*[2]

Because the use of force, f, requires energy from a built-in hazard to create the d in work, W, all work involves some level of risk. Occasionally, people lose control of these hazards. Human fallibility is an inherent characteristic of the human condition—it's in our genes. Error is normal—a fact of life, a natural part of being human. The human tendency to err is not a problem until it occurs in sync with significant transfers of energy, movements of matter, or transmissions of information.[3] In an operational environment, human error is better characterized as a *loss of control*—a human action that triggers an unintended and unwanted transfer of energy (ΔE), movement of matter (ΔM), or transmission of information (ΔI).[4] Human performance (**Hu**) is the greatest source of variation in any operation, and the uncertainty in human performance can never be eliminated. If work is not performed under control, the change (d) may not be what you want; work can inflict harm. *Work involves the use of force under the condition of uncertainty.*[5]

When performing work, people usually concentrate on accomplishing their immediate production goal, not necessarily on safety.[6] Most of the time, people's attention is on the work. If people cannot fully concentrate on being safe, thoroughly convinced there will be no unintended consequences 100 percent of the time, then *when* should they fully focus on safe outcomes?

VALUE ADDED VERSUS VALUE EXTRACTED

Recall Dorian Conger's quote at the beginning of this chapter, "If an operation has the capacity to do work, it has the capacity to do harm." All human-directed work intends to accomplish something that meets customer requirements, to add value.

---

* Work is the application of physical strength or mental effort to achieve a desired result, whether a force over a distance or careful reasoning (still a force over distance, though at a microscopic level).

However, when people manipulate the controls of built-in hazards, there is a corresponding risk to do harm that can extract value instead of adding value. The greater the amount of energy transferred, matter transported, or sensitive information transmitted during a human action, the greater the potential harm. Those human actions or procedure steps that can trigger serious harm must go right the first time every time. If the severity of harm potentially suffered by an asset would be considered *intolerable*, that action would be considered a CRITICAL STEP. An event/incident/accident is a form of value extraction.[7]

Referring to Figure 1.1, work may involve interactions with several assets, in this case two. At least two assets are in play for every work activity: typically, the person doing the work from a personal safety perspective and the product of their work from a business perspective. Figure 1.1 illustrates that steps 8 and 12 involve interactions with asset 1, and steps 4 and 17 require interactions with asset 2. If the performer loses control of the work at step 4, while doing work on asset 2, the amount of work done at that step would not trigger enough harm to the asset to exceed the degree of harm deemed intolerable, albeit some harm ensues. However, if the performer loses control at step 17, again working with asset 2, it is possible for asset 2 to suffer sufficient harm that would exceed a level of severity that managers previously deemed intolerable. The same logic applies to work on asset 1. Consider the following points to better understand the illustration:

- The horizontal line (x-axis) represents steps (denoted by dots) in a work activity, whether directed by a procedure or skill-of-the-craft.

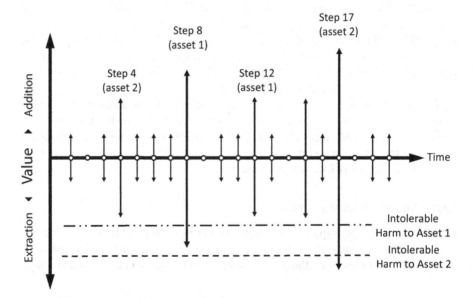

FIGURE 1.1 Value added versus value extracted (harm) during work. If an operation has the capacity to do work, it has the capacity to do harm.

- The vertical axis (y-axis) represents value added (above the horizontal line) or value extracted (below the horizontal line).
- The length of vertical lines denotes the degree of value (work done), either added or extracted.
- Intolerable harm is denoted by horizontal dashed lines for each asset.
- Some steps/actions involve one or more assets, or none.

The degree of intolerable harm to a particular asset that it could potentially suffer should be understood before work (usually a management decision) to designate a particular step as a CRITICAL STEP.

> **Note**: Not every action that has a point of no return is a CRITICAL STEP. Just because an action cannot be undone does not, by itself, constitute a CRITICAL STEP. The designation of an action as a CRITICAL STEP depends primarily on the *degree of harm* experienced after a loss of control of that specific action, what managers consider *intolerable*.

On that fateful night, the Colonel was uncertain as to the identity of the two vehicles thought to be enemy. He initially attempted to engage the vehicles with the Apache's machine guns, which would have inflicted less damage than a missile, less likely killing anyone inside. Less energy, less harm. But those guns jammed.

CRITICAL STEP DEFINED

DuPont de Nemours, Inc., commonly known as DuPont, defines *operational discipline* (or OD) as, "the deeply rooted dedication and commitment by every member of an organization to carry out each task the right way every time. Do *it* right the first time, every time."[8] But, *it* can be almost anything—*it* must be more specifically defined. We call *it* a CRITICAL STEP. Does every human action have to be performed perfectly? Let's conduct a simple thought experiment.

Thirty-year-old Jill arrives for work well rested after a good night's sleep. She's conscientious, enjoys good health, and has strong family support. Personal problems do not weigh her down. In short, Jill is well-trained, mentally alert, and physically fit and faces minimal emotional distractions—an ideal worker. For illustration purposes, let's assume that Jill is 99 percent reliable* for the task she is given when she arrives at work.[9]

Jill's supervisor assigns her a task that consists of exactly 100 actions. Let's assume the working conditions for every action are the same throughout the job—the chance for success is the same for step 100 as it is for step 1. Here's the question. What is the likelihood that Jill will perform *all* 100 actions without error?

---

* Reliability is the likelihood of successful performance of a function.

Jill's performance is a simple probability calculation. The chance for success on step 1 is 0.99; the chance for success in step 2 is the same, 0.99; and the chance for success on step 3 is—you guessed it—0.99, and so on to the 100th action. The mathematical equation for the probability of successfully completing *all 100 actions without losing control* is:

$$p100 = 0.99 \times 0.99 \times 0.99 \cdots 0.99^{100} \cong 0.3660 \; or \approx \mathbf{37\%}$$

It may astound you that the chance of performing just 100 actions without error is only 37 percent for someone who is 99 percent reliable.[10] There's a much better than 50-50 chance that Jill will do something wrong along the way (63 percent). The news is better if a person's reliability is near the top of the nominal human reliability scale—99.9 percent; but even then, the probability of successfully completing all 100 actions without losing control improves to just 90 percent. That still equates to a 1 in 10 chance of erring at some point in the 100-step task. A mistake at some steps may not matter. For most work, 99 percent reliability is acceptable. The question to ask is, "Which action absolutely has to go right the first time, every time?" These are the points in the task at hand, which Jill and her boss should identify for her to success-fully complete the task without experiencing serious injury, loss, or damage.[11] So, let's restate our definition:

> A CRITICAL STEP is a human action that will trigger immediate, irreversible, and intolerable harm to an asset if that action or a preceding action is performed improperly.[12]

Battlefields are complex adaptive systems that breed ambiguity, uncertainty, and volatility—otherwise known as a VUCA environment.[13] Misreading the grid coordinates to the alleged targets and misunderstanding friendly forces to be enemy, compelled by ominous radar alerts, his fellow pilots, and the ground commander, the Colonel let loose two missiles—the CRITICAL STEP. Once launched, the missiles were beyond control—they would follow the physics of their design and the environment, eventually fulfilling their deadly purpose. The Colonel and his copilot experienced VUCA and got things wrong, albeit unintentionally.

MAINTENANCE TEST GONE WRONG[14]

An operating furnace was inadvertently shut down during preventive mainte-nance on a safety-related instrument.

Each of two boilers was fitted with a temperature recorder-controller and a respective high-temperature trip function. The two recorders were positioned side by side on the front of the control room instrument panel, with A recorder on the left and B recorder on the right, as shown in Figure 1.2.

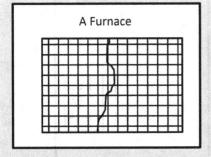
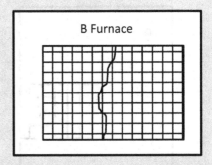

FIGURE 1.2 The view of the two strip chart recorders on the control panel, as seen when standing in the control room. Each recorder also functioned as a controller for its respective furnace.

An instrument technician was directed to check the calibration of the high-temperature trip feature on A furnace (a combined recorder-controller). The technician placed the controller in manual* and then walked behind the control panel to access the rear of the recorder-controllers. The next several steps of the procedure were to 1) remove the cover from a junction box, 2) disconnect (lift) the signal lead (electrical wire from detector) from the furnace temperature detector, 3) connect a test box to the controller, 4) apply a gradually increasing temperature signal from the test box, and 5) note the reading on the recorder at which the trip would occur.

Behind the control panel, the junction boxes for A and B are in line with the recorders (front to back). Therefore, when viewed from behind the control panel, looking toward the control room proper, the B recorder-controller junction box is on the left as shown in Figure 1.3.

There was a label for each recorder-controller in the rear access area; but it was near the floor, not on the junction boxes, and the font was small. This was not necessarily a major factor in this incident because this technician had done this task several times before. Better labeling could have caught the attention of the technician before the fateful action. Remember, the A recorder/controller is in manual; the B recorder/controller is still active.

The technician removes the cover from the B junction box and disconnects the signal lead. Bang! The B furnace trips, shutting down. The A furnace is still operating. The effect of disconnecting the signal lead is the same as a failed temperature detector, which correlates to a maximum high temperature. The controller signals the fuel supply valve to close, and the furnace shuts down.

---

* Placing the recorder/controller in manual bypasses the trip feature from the controller such that the temperature signal for furnace A is locked at the current temperature, and the furnace's fuel-supply valve position stops being driven by the controller.

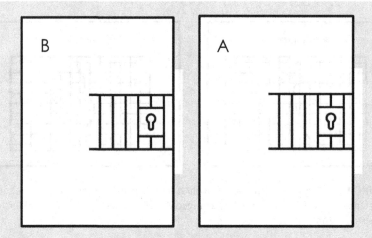

FIGURE 1.3 The view of the junction boxes as seen from behind the control panel. The instrument technician was required to unlock the junction box to access the signal wire to the recorder-controller. No labels were on the rear faces of the junction boxes—the letters B and A are noted to let the reader know that from the perspective of the technician from behind the control panel, the B junction box is on the left and the A junction box is on the technician's right.

We hope you recognized the CRITICAL STEP—lifting the lead from the B furnace. Question: The harm is an interruption of services provided by the B furnace. Does disconnecting the lead satisfy the definition of a CRITICAL STEP? The asset is the furnace and the service it provides, and the hazard is an interruption of fuel.

- Will the human action result in a change in state of the furnace? Yes—shut it down by closing the fuel supply valve.
- Is the change in state immediate? Yes—mere moments, faster than humans can respond to avoid the consequence.
- Is the change in state irreversible? Yes—the furnace shuts down, terminating services for a period of time.
- Is the change in state harmful and intolerable? Presumably, yes. If the furnace was supplying vital services, it may involve serious losses or damage to customers or other assets. (Management must decide what level of harm is considered intolerable.)

Would lifting the signal wire to A furnace recorder-controller be considered a CRITICAL STEP, even if the channel is in manual? Yes! Both actions to lift the signal lead are critical to the operation of their respective furnaces. The A furnace would not trip because its respective recorder/controller was in manual, while the B furnace controller was still active ("in control"). At the conclusion of Chapter 5, we reveal that CRITICAL STEPS are *always* considered critical regardless of absence of a hazard. Chapter 4 will do a deep dive on *Risk-Important Actions* (RIAs). Placing the controller in manual was an RIA for the CRITICAL STEP of disconnecting the signal lead, allowing the recorder-controller to respond to a simulated signal without triggering an inadvertent protective control action.

Would you recognize a CRITICAL STEP if you saw one? What criteria would you use to conclude a procedure step or other human action is a CRITICAL STEP? To help you accurately identify a CRITICAL STEP, it is necessary to know its attributes, which are embedded in the definition.

ATTRIBUTES OF A CRITICAL STEP

Notice that the central idea in the definition of a CRITICAL STEP is the degree of *harm* to something of importance—an asset. An asset is anything of substantial or inherent value to an organization, such as people, property, product, and even productivity. Other factors are important to what actions would be considered CRITICAL STEPS. But without "intolerable harm," a human action that satisfies all other attributes would not be a CRITICAL STEP. A CRITICAL STEP's attributes, which are described in Table 1.1, are derived from Dr. David Embry's work in human reliability analysis (HRA). His research provides insight into the development of these attributes.[15]

A methodology for identifying a critical task was first developed by Dr. Embrey in a 1994 book he wrote for the American Institute of Chemical Engineers, *Guidelines for Preventing Human Error in Process Safety*. Dr. Embrey developed a framework for evaluating human sources of risk in an operating plant known as System for Predictive Error Analysis and Reduction (SPEAR). The SPEAR methodology focused on tasks with significant risk potential, identifying human errors. We borrow and expand on the concept of a critical task by focusing not on errors but on harm to assets. Although the term CRITICAL STEP was borrowed from the DOE, the principles and practices of CRITICAL STEPS introduced in this book springboard off the SPEAR logic, which was to identify the human interactions with a system that would have adverse impact on risk if errors occured. The screening process of the SPEAR methodology asked the following questions:

1. Is a hazard present in the area of the operation? (Potential to cause harm—severity)
2. Given that a hazard is present, could any human interactions cause harm? (Hands-on manipulations by people who could trigger release of the hazard)
3. Given that workers interact with hazardous systems, how frequently would they err in this critical task? (Likelihood of error; i.e., losing control of the hazard)

TABLE 1.1
Attributes of CRITICAL STEPS.

Attribute	Description	Common Examples
Human	Hands-on performance by frontline workers; persons in direct contact with assets or control of related hazards during operations	• Operator/electrician/craftsman • Nurse/surgeon • Pilot • Information technology (IT) tech Not: equipment, knowledge-worker
Action	Physical activity by people that involves hands-on exertion of a force on an object (act of commission)	• Push/pull/lift/turn/flip • Handle/tap/punch • Walk/run/kick/nudge • Depress (enter) Not: think, decide, speak
Will	• Certainty; complete assurance that energy will be transferred, matter will move from one place to another, or information will be transmitted • Assurance of the onset of harm to an asset	• Unavoidable burn after touching hot stove • Inescapable after stepping into a bear trap • Irrevocable after pulling a fire alarm Not: maybe, likely, could
Immediate	Faster than a human can react or respond to avoid consequences	• Instantly (explosion/spark) • Split second (spill/crack) • Moments (loss of cooling flow) • Seconds to minutes (overheating) Not: delays, hours, days, weeks
Irreversible	• One or more critical parameters of an asset are exceeded, resulting in permanent change • Past the point of no return • No undo—the onset of harm is inevitable • Inability to reestablish conditions prior to action	• Brain damage (and intolerable) • Burned toast (not intolerable) • Un-ringing a bell • Returning bullet to a firearm's chamber after shooting it Not: irrecoverable (equipment damage is recoverable at a cost; human life is not)
Intolerable harm	• Disabling injury or death • Significant damage or substantial loss • Defined for every asset • Severity of harm meets organization's definition of an event; reportable to a regulator • Severity defined by what regulatory agencies consider unacceptable • Dependent on what the managers consider important to safety, quality, the environment, production, etc.	• Death/permanent disability • Severed limb • Damage/cost exceeding $50,000 • Unacceptable quality to a customer • Loss of mission functionality • Loss of market share/out of business Not: paper cut, embarrassment, minor water spill, $50 cost

Note: The goal is to help identify human actions or activities that pose the greatest risk to an organization's assets during production operations. Consequently, definitions of words used to define a CRITICAL STEP are important to understand.

Answers to the SPEAR @jim@jamaroperations.com @ron@hopeconsultingllc.com questions are used to rank the risk potential of various work activities for a more detailed HRA, which is not within the scope of this book. In 1997, Dr. James Reason made a practical observation about this process in his book *Managing the Risks of Organizational Accidents*. He mentioned that non-specialists in HRA, such as procedure writers, quality inspectors, and managers, should pay less attention to human error and more on the *consequences* of it to the system and its products.[16] We agree. CSM, which is described in Chapter 7, provides a similar analytical method for systematically identifying perpetual CRITICAL STEPS in operational processes and procedures. This is our focus.

> **Caution**: On the surface, the concept of CRITICAL STEP appears to be simple and straightforward. The mistake made most often is that too many actions or steps are considered *critical*. People tend to conflate RIAs with CRITICAL STEPS, which strongly suggests they haven't internalized the attributes of a genuine CRITICAL STEP. As the old saying goes, "If everything is important, then nothing is important."

To be useful in managing human performance risk, the concept of a CRITICAL STEP must be reserved for 1) what is vital to the life and health of workers and the public, 2) the essential functioning of safety-critical plant equipment, 3) the quality of goods and services delivered to customers, and ultimately, 4) the economic survival of the organization. Frontline workers, procedure writers, supervisors, and others who do hands-on work must discipline their use of CRITICAL STEPS in operations.

The following list offers a few everyday examples of human actions that satisfy the definition and criteria of a CRITICAL STEP. You might be surprised by a few.

- Depressing the "Trip" pushbutton on a circuit breaker's physical control panel that supplies electric power to a hospital
- Walking through the opening into a confined space (possible oxygen-deficient atmosphere)
- Making an incision on a patient during surgery
- Turning on the kitchen sink garbage disposal unit
- Pulling a fuse or an integrated circuit (IC) card from a digital control system
- Grasping a bare electrical cable or wire
- Clicking "Send," "Submit," "Start," or depressing the "Enter" key
- Loosening bolts on a pipe flange or manway cover on a high-pressure system
- Touching the shaft of an operating pump (rotating at 1,800 rpm) with your hand
- Extracting a tooth
- Leaping across the open door of an airborne aircraft while skydiving
- Depressing the accelerator of your automobile
- Walking across a street

In every example, there is a human action (verbs ending with "ing"), and a transfer (or interruption) of energy (electrical, mechanical, heat, etc.), a movement of matter (solid, liquid, or gas), or a transmission of information (data, information, software, signals, authorizations, etc.) that could trigger immediate, irreversible, intolerable harm to something important.

TECHNICAL EXPERTISE—THE BEDROCK OF
CRITICAL STEPS AND RISK-BASED THINKING

When an asset suffers serious harm, the boundaries of what is safe for the asset were exceeded. If frontline workers are to protect assets from harm during everyday work, they must possess the prerequisite technical knowledge and understanding of the safety boundaries for all the assets they work with on the job. To exercise RISK-BASED THINKING, the presumption is that the person understands the technology. Otherwise, how could a person *know* what to anticipate, *know* what to monitor, or *know* what to do to exercise positive control and to protect assets?

Because of their expertise and humility, the best performers have a *deep-rooted respect* for the technology as well as their own fallibility. Expertise is more than technical knowledge. Expertise includes understanding, experience, and proficiency. Practitioners, operators, and craftsmen not only understand the safe and proper means for transfers of energy, movements of matter, or transmissions of information; they also understand the when and how pathways between built-in hazards and assets are created and how they could trigger harm—if they lose control. Top performers continuously update their awareness of hazards and their proximity to assets. Consequently, they more readily anticipate the worst, recognize the mistakes they dare not make, and equip themselves to respond appropriately.[17] This level of knowledge and skill becomes a key ingredient to "expert intuition," which will be discussed in Chapter 2.

> **Caution:** Some line managers ascended to their positions, not because of their technical background, but due to their administrative skills. While most line managers don't need to possess the same level of technical expertise as their subordinate frontline workers, manager also must possess a deep-rooted respect for the technology.

But, in the long run, expertise applied without the input and corroboration of other competent persons is more vulnerable to error, increasing the workers' potential to lose control—hence, the importance of group conversations characterized with robust dialogues that reveal the technical realities of high-risk work. Recent experience shows that technical expertise practiced collectively is more powerful than when it is practiced individually.[18]

CRITICAL STEPS IMPROVE EFFICIENCY

Identifying and controlling CRITICAL STEPS help you navigate the safety/production space, optimizing the use of already scarce safety resources. CRITICAL STEPS improve the efficiency of human performance by highlighting those human actions, steps, or phases of work that must go right the first time, every time. Trade-offs between efficiency and safety are normal and occasionally necessary to meet deadlines. You cannot always be utterly thorough from a safety perspective and still stay competitive. It's inefficient (and impossible) to attempt to prevent human error on every step and human action of every operation. And you cannot operate with 100 percent efficiency, because some resources are redirected toward safety functions—controls, barriers, and safeguards.[19] But you always want to meet your commercial deadlines, safely, with the required quality committments. You have to navigate a middle ground to accomplish both safety and profitability goals during work.

For complicated operations, procedures prescribe a series of human actions organized in a preferred sequence to accomplish production and safety goals. But which actions require vigilance and heightened attention—*which human actions absolutely must go right*? Most human actions in an operation can be described as non-critical. At those times, it may be acceptable to err on the side of efficiency, considering nominal human reliability. Good enough (99.9 percent) is good enough when nothing is at stake. Recall Figure 1.1 and Jill's 100-step task thought experiment. For low-risk operations, we believe it is acceptable to speed things up to reduce costs.

Incorporating the principles and practices of managing Critical Steps into your operations has as much to do with efficiency and productivity as with safety. You have the option to expedite those portions of a task that have little to no risk to safety and the business, but you must absolutely slow down for those that do. Slowing down should be deliberate, guided by a persistent chronic unease,* conversations, and Risk-Based Thinking. By isolating the more relevant and important human risks, identifying and exercising positive control of Critical Steps enhance both safety (thoroughness) and productivity (efficiency).[20]

EXCELLENCE IS NOT GOOD ENOUGH!

Excellence is always described in relative terms as possessing an outstanding quality or superior merit—remarkably good, *compared to others*. People do things right most of the time. Nominal human reliability drifts between 99 and 99.9 percent, depending on local factors.[21] Is 99 percent good enough? Furthermore, is 99.9 percent good enough for Critical Steps? We think not. As the Colonel experienced, sometimes doing your human best is not good enough.

But when it comes to doing the right thing and doing the right thing right, such as a Critical Step, it becomes imperative to avoid losing control. When nothing significant is at stake, 99 percent is satisfactory. This performance level is fine if the person is simply taking care of household chores—making the bed, brushing one's teeth, setting the table, placing dishes in the dishwasher, vacuuming the carpet, painting a bedroom, or reading a book. Around the house, for instance, most people rarely experience a genuine problem by performing at that rate of human reliability.

On the contrary, precision and accuracy in execution are more important than speed in high-risk performance situations.[22] High-risk work MUST slow down to allow frontline workers to think and act mindfully—to be deliberately certain that assets are

---

* Generally, the experience of concern about risks, exemplified by a healthy skepticism about one's decisions and the risks inherent in work environments. Operationally, an ongoing wariness of hidden threats in the workplace that could trigger harm, spawned by a deep-rooted respect for the technology, its complexities, and its built-in hazards.

protected from harm despite production pressures. Therefore, it is strategically essential to define and understand what must absolutely go right; without doing so, the cost of failure lies in the harm to key assets. The business case is self-evident. Managers, engineers, supervisors, and workers must all know, understand, and agree on what must go right, especially during high-tempo operations that experience schedule and budget pressures. It is important to recognize that bias toward speed and efficiency is NEVER appropriate at CRITICAL STEPS and RIAs. The risk is simply too great. When a loss of control must be avoided, **precision execution is the ONLY acceptable standard!**

KEY TAKEAWAYS

1. A CRITICAL STEP is a single human action that will trigger immediate, irreversible, and intolerable harm to an asset if that action or a preceding action is performed improperly.
2. If an operation has the capacity to do work, then it has the capacity to do harm. Work is energy directed by human beings to create value. Therefore, work involves the use of force under conditions of uncertainty—that is, risk.
3. The central attribute in the definition of a CRITICAL STEP is the degree of *harm* (intolerable) to something of importance—an asset.
4. To be useful in managing human performance risk, the concept of a CRITICAL STEP must be reserved for what is profoundly important to safety, quality, reliability, and productivity.
5. CRITICAL STEPS improve efficiency of human performance by highlighting those steps or phases of work that absolutely must go right. All other portions of the task that have little to no risk to safety, quality, reliability, and productivity may be performed with deference to efficiencies, while maintaining a mindset of chronic unease.
6. A comprehensive understanding of the technology and its hazards—a deeprooted respect—is necessary to reliably recognize CRITICAL STEPS.
7. Excellence is not good enough at CRITICAL STEPS. Precision execution is the only acceptable performance standard.

CHECKS FOR UNDERSTANDING

1. Which of the following actions with a handgun is a CRITICAL STEP?
   a. Loading the firearm with bullets
   b. Cocking the firearm—pulling the hammer back
   c. Pointing the muzzle at a target
   d. Moving the safety lever off SAFE
   e. Pulling the trigger
2. Which attribute of a CRITICAL STEP is most important?
   a. The harm is irreversible.
   b. The harm is immediate.
   c. The harm is intolerable.

3. True or False. Donning safety equipment, such as hardhats, eye and ear protection, gloves, is a CRITICAL STEP.

(See Appendix 3 for answers.)

THINGS YOU CAN DO TOMORROW

1. Develop an operational definition of a CRITICAL STEP that is relevant to each work groups' work. Verify it satisfies all the attributes of a CRITICAL STEP.
2. Print several posters with the CRITICAL STEP definition. Display them prominently, able to be read from across a room, in work areas and production meeting rooms, especially where prework discussions would occur, and in training settings.
3. Identify operations or tasks currently scheduled that are high-risk or potentially costly if control is lost. Using your definition, pinpoint CRITICAL STEPS and the potential harm to assets. Explore means to exercise positive control and to limit harm (to fail safely).
4. Using a previous human performance event as a case study, ask your work group to identify the CRITICAL STEPS using this book's definition. Ask the group to judge whether the proposed CRITICAL STEPS satisfy the definition.
5. Just before performing high-risk work activities, ask workers to pinpoint those one, two, or three actions that must absolutely go right the first time, every time. Compare those actions with the definition of a CRITICAL STEP.

REFERENCES

1 U.S. General Accounting Office (1983, June). *Operation Desert Storm—Apache Helicopter Fratricide Incident, Report to the Chairman, Subcommittee on Oversight and Investigations,* Committee on Energy and Commerce, House of Representatives (GAO/OSI-93-4).
2 Muschara, T. (2018). *Risk-Based Thinking: Managing the Uncertainty of Human Error in Operations.* New York: Routledge (p. 24).
3 Center for Chemical Process Safety (1994). *Guidelines for Preventing Human Error in Process Safety.* New York: American Institute of Chemical Engineers (pp. 207–211).
4 Hollnagel, E. (2004). *Barriers and Accident Prevention.* Burlington: Ashgate (pp. 76–78).
5 Muschara, T. (2018). *Risk-Based Thinking: Managing the Uncertainty of Human Error in Operations.* New York: Routledge (p. 25).
6 Hollnagel, E. (2009). 'The Four Cornerstones of Resilience Engineering.' In: Nemeth, C., Hollnagel, E., and Dekker, S. (eds.). *Resilience Engineering Perspectives Volume 2, Preparation and Restoration.* Farnham: Ashgate (p. 29).
7 Reason, J. (1997). *Managing the Risks of Organizational Accidents.* Farnham: Ashgate (p. 2).
8 Rains, B. (2010). *Operational Discipline: Does Your Organization Do the Job Right Every Time?* Wilmington, DE: DuPont Sustainable Solutions.
9 Nominal reliability rate derived from general human error rates described in Kletz, T. (2001). *An Engineer's View of Human Error* (3rd ed.). Boca Raton: CRC Press (pp. 138–139).
10 Crosby, P. (1984). *Quality Without Tears.* New York: McGraw-Hill (p. 76).
11 Muschara, T. (2018). *Risk-Based Thinking: Managing the Uncertainty of Human Error in Operations.* New York: Routledge (p. 103).

12 Ibid. (p. 271).
13 Barber, Herbert F. (1992). 'Developing Strategic Leadership: The US Army War College Experience.' *Journal of Management Development*, 11(6): 4–12.
14 This incident is adapted from Kletz, T. (1994). *What Went Wrong? Case Histories of Process Plant Disasters* (3rd ed.). Houston: Gulf (pp. 69–70).
15 Center for Chemical Process Safety (CCPS) (1994). *Guidelines for Preventing Human Error in Process Safety*. New York: American Institute of Chemical Engineers (pp. 207–211).
16 Reason, J. (1997). *Managing the Risks of Organizational Accidents*. Burlington: Ashgate (p. 91).
17 Weick, K., and Sutcliffe, K. (2007). *Managing the Unexpected: Resilient Performance in an Age of Uncertainty* (2nd ed.). San Francisco: Jossey-Bass (p. 46).
18 McChrystal, S. (2015). *Team of Teams: New Rules of Engagement for a Complex World*. New York: Portfolio/Penguin (pp. 167–169).
19 Reason, J. (1997). *Managing the Risks of Organizational Accidents*. Burlington: Ashgate (p. 28). It's also notable that the second law of thermodynamics means there is always some energy wasted in a process.
20 Hollnagel, E. (2009). *The ETTO Principle: Efficiency-Thoroughness Trade-Off: Why Things That Go Right Sometimes Go Wrong*. Burlington: Ashgate (pp. 25–30).
21 Swain, A., and Guttmann, H. (1983). *Handbook of Human Reliability Analysis with Emphasis on Nuclear Power Plant Applications: Final Report* (NUREG/CR-1278). Washington, DC: U.S. Nuclear Regulatory Commission.
22 Hollnagel, E. (2009). *The ETTO Principle: Efficiency-Thoroughness Trade-Off: Why Things That Go Right Sometimes Go Wrong*. Farnham: Ashgate (p. 52).

2 Thinking about Human Performance Risk

You cannot change the human condition, but you can change the conditions in which humans work.[1]

—Dr. James Reason Ph.D. and Professor Emeritus

What does this button do?

—Unknown

LIFE CHANGING STEP

As a seasoned nuclear navy operator in the late 1980s, I (Ron) had spent my last 3 years of an 8-year career training enlisted and officer nuclear operator trainees at a prototype training facility in Idaho. Most of my days were spent running drills, checking trainees' knowledge on power plant systems, and sitting on qualification boards to evaluate each trainee's readiness for the rigor of life aboard U.S. Navy nuclear-powered vessels.

Early in my assignment, I qualified to be one of a handful of gas-free engineers.* One evening, I was directed to check the atmosphere of a large, empty, 30-foot deep, 40-foot-wide tank that typically held water. This task required me to check the tank's atmosphere with various test devices after it had been purged with fresh air. This was common practice, and on any given shift I might be asked to test several empty tanks and voids. This testing was done to verify that the tank's atmosphere was safe for human habitation. Typically, I would notify the control room before each tank entry—no formal prework discussion was conducted even though the task is inherently dangerous.

In those days, the Navy had limited fall protection requirements for ascending and descending ladders. There was no clear requirement to be tied off, and there was no fall arrest equipment for that matter, as is required and used today. My only required personal protective equipment (PPE) for this entry included coveralls, a double layer of outer rubber gloves with thin cotton inner gloves, double plastic booties (notably slick), and a self-contained breathing

---

* The gas-free engineer is qualified to certify a confined space as being safe for others to enter without the use of an air-purifying or supplied air respirator (SAR). A confined space would, however, need to be ventilated prior to entry to ensure an adequate supply of breathable air. Gas-free engineering is equivalent to a toxic gas inspector, confined space inspector, etc.

DOI: 10.1201/9781003220213-2

apparatus (SCBA) strapped to my back, which hampered my visibility and maneuverability.

After putting on the "proper" PPE, I performed some initial tests at the opening of the manway and found no abnormal gas readings. I hung my test equipment around my neck so I could descend into the tank. As I descended the ladder into the tank, maintaining three-point contact, three rungs from the top one foot slipped between two rungs. I fell backward. I frantically reached out for the ladder rungs, grasping only air, when suddenly and luckily the handwheel of the tank of my SCBA caught the edge of the access opening behind me just long enough for me to grab the ladder. Breathing heavily, I held myself in place for a few moments while I gathered my wits. I quickly exited the tank and laid prostrate on the tank roof, with my heart pounding out of my chest. Several minutes passed before I could move. I am not sure that I had ever been more scared than I was for those few seconds when I thought I was going to die. I removed my PPE and reported the incident to my supervisor.

Does Ron's descent into the tank on the ladder satisfy the definition of a CRITICAL STEP? Consider the following facts:

- He was physically poised 30 feet above the bottom of the tank (pathway).
- Each step on the ladder rungs was a human action, subject to his own fallibility.
- He was wearing slick rubber boots not conducive to secure traction on the ladder's rungs when his feet slipped.
- The sudden—immediate—slip jostled Ron such that he lost his grip of the ladder, falling backward away from the ladder due to the weight of the SCBA strapped to his back.
- Fortunately, the air tank valve caught the edge of the access (luck).
- It would take only a few moments to fall 30 feet to the floor.
- He was not wearing a fall-protection safety harness, which was not a required safety device at the time (impacts ability to "fail safely").
- The impact with the tank floor would either have killed Ron or at least broken some limbs or caused internal injuries—all intolerable.

The answer is yes. All the attributes of a CRITICAL STEP's definition are present: improper human action and immediate, irreversible, intolerable harm (a near hit). In fact, the first step onto the ladder was the first CRITICAL STEP of many (literally, in this case). One slip—one misstep—almost cost Ron's life. Fortunately for Ron, he lived to tell the story—this story. He was lucky. But, as we should all know, luck is not a reliable defense. Yet he did everything right according to the Navy's safety standards at the time. *He believed he was safe.* This incident became a defining moment for Ron—truly, believing it was an act of God that he survived; others call it coincidence. While he was more than a few feet or so above the floor of the tank, each step descending the ladder was "critical"; every step had to be performed with precision. You could say Ron's descent on the ladder was a "continuous CRITICAL STEP."

HUMAN ERROR = LOSS OF CONTROL

Humans are a key source of variation in operations (uncertainty due to fallibility). Of all the activities of an operation, human performance is the least reliable. "Human error" is often simply an action inappropriate for current conditions. Regardless of what it's called or labeled, human error, active error* in particular, is a principal source of risk to the assets of production and safety. That includes people. Yes, people can be a hazard, but they are also heroes. People, especially those in the workplace, are also a key source of resilience because of their adaptive capacity. It is our systems and people together that lead to successful outcomes. This concept will be developed more fully later when we talk about augmenting adaptive capacity.

An event is an undesirable occurrence involving significant harm (injury, damage, or loss) to one or more assets due to an *uncontrolled* transfer of energy, movement of matter, or transmission of information. The anchor point in any event occurs at a point in time when control over the damaging properties of energy, matter, and information is lost—when the destructive potential of built-in hazards is unleashed because of a loss of control and/or the absence of adequate protection.[2]

When errors trigger events, too often people are blamed for their lack of judgment or carelessness. We encourage the reader to think of human error not as some immoral act—"They should have known better"—but to think of it as a *loss of control*. Senior managers simply (and immorally) blame the individual, ignoring relevant system defects. This happened to Ron. He did everything "right." But the Navy system was not designed to accommodate human fallibility. This chapter describes a way of thinking about the risk human performance poses to assets in the workplace, another type of battlefield. Better thinking yields better management.

HUMAN PERFORMANCE RISK CONCEPT

To manage or control anything, you must understand how it works. Mental models are representations of how things work, explanations of cause and effect that managers commonly use to make decisions. Good mental models allow managers to see relationships, to ask better questions, and to predict outcomes of a decision or action more reliably and accurately. When working with complex adaptive systems involving multiple human tasks, relationships between system components become obscure. Therefore, interactions and the effects of feedback must be checked periodically to validate the mental model's reliability and utility. Consequently, it is always important to cling to a *sense of unease* when using mental models to make decisions about safety. We are about to introduce a couple of them to you.

---

* Active errors are those occasions when a human action triggers immediate harm.

Whenever work is done to create value, three physical things are present concurrently[3]:

- *Assets*—things important, of high value, to an organization
- *Hazards*—built-in sources of energy, matter, or information used for work to create value
- *Human beings*—actions by fallible people intending to create value

Arranging these elements into a conceptual model, we develop a more systematic way of thinking about risk. This model helps you think about and manage the risk introduced by the uncertainty of human performance. Figure 2.1 illustrates the occurrences in a work activity, where all three are in intimate proximity or physical contact, generating risk—risk to the asset.[4] This risk is not usually a permanent state of work. Most human activity involves set-up, communications, adjustments, clean-up, etc.; the coupling of the three elements of risk occurs less often during operations, except where production work is performed.

CRITICAL STEPS are to be performed with the same sense of unease as with the use of a firearm. Either way, you can end up dead. Although the use of firearms may not be familiar to everyone reading this book, anyone who has watched an action movie has seen firearms used. Hollywood loves firearms. Take a handgun with a cartridge (bullet) in the chamber—a hazard to any living thing. The safety is off. The firearm is ready to shoot. The person with a scowl on the face, wielding the firearm, has a finger on the trigger, and the muzzle is pointed at another human being. Tension rises because there is now risk.

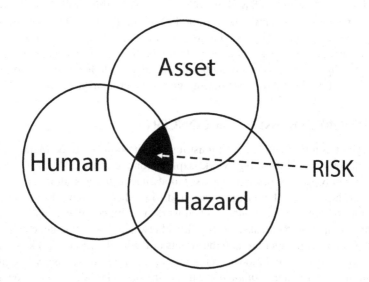

FIGURE 2.1   The *Hu Risk Concept* illustrates the primary elements in managing the risk that human fallibility poses to operations. The interfaces (overlaps) of these elements introduce risk: losing control of a built-in hazard and harming an asset—an event.

Source: See Muschara, T. (2018). *Risk-Based Thinking* (p. 26).

As illustrated, the **Hu Risk Concept** is a control problem: control of 1) the co-location of a hazard with an asset, 2) human interactions with either, or 3) the moderation of built-in hazards used during work processes.[5] A means of controlling (managing) human performance risk is described in the next section. But, first, let's better understand the three elements that create risk.

Asset—Things of Value to Protect from Harm

Assets include anything of value, tangible and intangible, important to the organization's reason for being—its mission. For a business or organization to be sustainable, the assets, such as people, product, property, facilities, equipment, and even shipping labels, used to create or deliver a company's outputs must be protected from harm. Whatever is essential or key to its safety, productivity, reliability, environment, and profitability is of utmost importance to the members of a responsible organization.

Harm is defined by the asset, where permanent damage, injury, or loss can be sustained. All tangible assets have a safe operating envelope (SOE), defined by one or more critical parameters, within which the asset's integrity (safety) is preserved, if the respective critical parameters are not exceeded.[6] For example, the SOE of a car tire includes wall condition, tread depth, air pressure, temperature, and speed.[7] The collection of critical parameters for one or more assets is frequently called the system's "design basis." Measures of an asset's critical parameters serve as the vital signs of the asset's safety, quality, system reliability, or operability of equipment and processes. In most cases, the conscientious use of procedures and built-in controls and barriers will help the frontline worker operate within an asset's SOE.

Hazards—Built-in Sources of Potential Harm

Work creates value and requires the transfer of energy, the movement of matter, or the creation and transmission of information. Work requires energy to create a change (recall, $W = f \cdot d$). Movements of matter require work. It takes work to create and communicate information. Hazards are built-in sources of potentially damaging energy, matter, and information, necessary for creating value during operations, research, services, etc.[8] All industrial facilities incur risks employing hazards in various forms to perform their functions, such as electrical energy needed to run a motor that drives a pump. See examples of the following three sources of hazards:

- *Energy*—electrical, kinetic, chemical, heat, elevation, thermal, such as ovens and stoves, hot surfaces, gasoline engines, nuclear reactors, rotating equipment
- *Matter*—transport of solids, liquids, or gases from one place to another, such as automobiles traveling on highways; seagoing tankers hauling large quantities of crude oil; pipelines carrying natural gas; pinch points; aircraft

traveling at 600 mph and at 35,000 feet; airborne viruses, bacteria, and various forms of contamination
- *Information*—data, documents, proprietary designs, trade secrets, instructions, policies, such as software, intellectual property, personal financial information

Regardless of the type of energy, all energy sources are hazardous when they exceed certain thresholds. Harm (damage, injury, or loss) is an unwanted change in the desirable qualities of an asset, defined by its critical safety parameters. Built-in physical hazards make harm a real possibility, especially when we lose control of them.[9]

We tend to assume hazards are stable—always present and knowable beforehand. Most are, some aren't. Occasionally, unknown hazards arise during work—landmines—appearing unexpectedly. Risk is dynamic; pathways for work (harm) from built-in hazards come and go. To sustain the safety of assets over the long term, workers must be capable of managing both known and unknown hazards (surprises) when they occur, that is, able to adapt and fail safely.

HUMAN FALLIBILITY—POTENTIAL FOR LOSING CONTROL

The occurrence of an event is usually triggered by some human action while a person is at the controls.[10] Variations in behavior, including human error, lead to variations in results. As described earlier, we encourage you to think of human error more as a loss of control than a fault of the individual. Human error is a normal feature of human nature—one is inhuman if faultless (most likely dead). Human performance introduces uncertainty at the exact time and place we want to create value. Nominal error of commission is approximately 3×10^{-3}, which equates to 99.7 percent reliability, roughly 1–3 errors in 1,000 attempts.[11] As mentioned previously, people are generally 99.9 percent reliable, ranging to 99.99 percent for expert performance. Sounds good, but would you rely on those numbers for life-and-death situations?

The guiding principle in the healthcare industry is "first, do no harm." That's the proper mindset that frontline workers, including their supervisors and managers, need in high-hazard operations. A risk-based, conceptual model of workplace human performance provides managers with the structured means to proactively manage human performance risk. On the contrary, an event-based approach—learning late by reactive reporting and event analysis—will only get you so far. So, what do you manage?

HUMAN PERFORMANCE RISK MANAGEMENT MODEL

The interfaces between hazard and asset, between human and hazard, and between human and asset (the overlaps in the *Hu Risk Concept* of Figure 2.1) provide opportunities for control. Referring to the *Hu Risk Concept* as a springboard, a more practical form, depicted in Figure 2.2, the *Hu Risk Management Model*, suggests what to manage—*pathways* and *touchpoints*. Pathways and touchpoints are necessary interactions to do work. But, if not managed

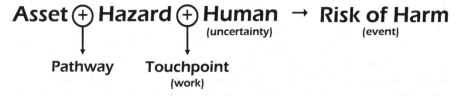

FIGURE 2.2 The *Hu Risk Management Model* pinpoints the two work-related interfaces—pathways and touchpoints (denoted by the plus (+) signs)—that must be controlled to minimize the risk of harm during operations.

Source: See Muschara, T. (2018). *Risk-Based Thinking* (p. 30).

proactively, harm can ensue during work with a loss of control. Interactions are denoted by plus signs (+).

> **Note**: Two types of human interactions are of interest in work: those with hazards and those with assets. This linear model is limited in that the interaction between human and asset is not represented explicitly, though the human-hazard interaction is the more important of the two. However, the touchpoint between the human and hazard is an adequate reminder that there is a touchpoint with the asset. All models are wrong, but some are useful.[12]

PATHWAYS

The first + sign represents the existence of a pathway for work, involving an impending interaction between an operational hazard and an asset.[13] Pathways are necessary for work to happen—a force is required to create a difference. A pathway exists when a hazard is poised in such a way as to *expose* an asset to the potential for a change in state—an opportunity for good (value) and a vulnerability for bad (threat). Whenever there is an opportunity to add value, there is an associated risk to do harm—to extract value. For example, the open door on an aircraft cruising at 13,000 feet offers a pathway for a skydiver about to jump. RIAs, performed earlier, create pathways. Risk matures—looms—at the creation of a pathway.

For Ron, the pathway was from the manway at the top of the tank to its floor 30 feet below. which was the only entrance into the tank. The ladder offered a "controlled" descent into the tank. Without fall protection, any slip would involve an irreversible fall after gravity takes over.

Pathways are particularly important because the potential for harm is now dependent on either a single human action or an equipment malfunction. For Ron, his descent was dependent on secure traction of every step. Frontline workers must be wary of the creation and existence of pathways, which occur often during production operations. But occasionally, pathways emerge unexpectedly, which we call landmines and will explore later in Chapter 5.

TOUCHPOINTS

The second + sign represents a human touchpoint. A touchpoint involves a human interaction with an asset or hazard that *changes the state* of that object through work. CRITICAL STEPS always occur at touchpoints. A touchpoint is work, involving a force applied to an asset or hazard over a distance, using tools or controls of hazardous processes. Manipulations occur at touchpoints that influence either the *status* of the asset and/or the *control* of a hazard. After a touchpoint is performed, things are different. Positive control of a touchpoint is most important when a pathway exists between an asset and a hazard. Because of human fallibility, risk occurs at touchpoints that involve a change in the state of assets; human error could occur during work or a loss of control of a hazard, such as a foot slipping off a ladder rung. Ron had four touchpoints as he descended the ladder: two feet and two hands. But he lost control of three of them; he didn't lose complete control because one foot remained on one rung. A touchpoint includes the following characteristics:

- *Human action*—bodily movements; exerting a force on anything
- *Interaction with an asset or hazard*—physical handling—force applied to an asset or hazard
- *Work*—force applied to an asset over distance or the control of a hazard*
- *Change in state*—"the result of work—distance" suggests changes in one or more parameters that define the state of the asset or the control of the hazard (*off* to *on*).

Table 2.1 offers some examples of important everyday tasks, pinpointing the associated assets, hazards, pathways, touchpoints, and potential harm if control is lost. The *responsiveness* and *consequence* of an interaction determine the importance of the human action involved. CRITICAL STEPS are those human actions that involve quick system responsiveness and severe consequences when control is lost. As described by Charles Perrow in his book *Normal Accidents*, a CRITICAL STEP is a good example of a "tight coupling" situation.[14]

> **Note**: As illustrated in Figure 2.1, *Hu Risk Concept*, risk exists when interfaces exist between an asset, a hazard that can do work, and a human that can influence either. When there is no pathway with a hazard, there is no risk to the asset—no work can be done. When there is a pathway but no touchpoints with either the asset or the hazard's control, there is no human performance risk. But it's the occurrence of touchpoints *after* the creation of a pathway that tends to be critical to safety. If a touchpoint is performed improperly, the performer can lose control, and harm—an event—is likely to occur.

Occasionally, the *human is the hazard*, as in the case of manual activities when people apply the force that triggers harm. In these situations, the pathway and the touchpoint are one and the same, such as a surgeon's cut, recording critical data,

---

* Thinking—the creation of knowledge—is work on a cellular level.

TABLE 2.1
Examples of Pathways and Touchpoints between Assets, Hazards, and Humans.

Task	Asset	Hazard	Pathway	Touchpoint	Harm
Crossing a street on foot	Pedestrian	Moving vehicular traffic	Street	Stepping into street	Death or bodily injury
Ironing (pressing) a garment	Delicate silk garment	Heat from iron	Iron poised inches above garment	Hand holding handle of iron	Permanently scorched garment
Shooting a firearm	Person	Bullet at high velocity	In front of muzzle	Finger on trigger	Death or bodily injury
Purchasing products or services on the Internet	Credit card number	Theft of identify	Internet connection (via Wi-Fi or cable)	Moving mouse pointer over "Buy Now" button or finger poised over *Enter* key	Loss of funds or personal financial information
Filling syringes with drug product via a filling machine	Drug product (liquid form)	Bacteria or other forms of contamination	Pipe or hose (with pressure differential)	Grasping valve handwheel	Contamination of drug product from upstream piping system
Family enjoying swimming at a neighborhood pool	Toddler (child)	Pool water	Airway to lungs	Standing within inches of edge of pool	Death from drowning
Reading e-mail messages	Personal computer (programs and data)	Malicious software (viruses)	Internet connection (via Wi-Fi or cable)	Finger poised over mouse with pointer over link	Loss of control of PC, loss of data, cost of recovery, delays, etc.
Cleaning a 480-volt circuit breaker	Electrician	High voltage	Metal conductor	Grasping metal wire	Shock/death from electrocution

chiseling a stone sculpture, sports in general, circus acts, and children between the ages of 2 and 5 (in jest, they must be corralled for their own protection). Shooting oneself in the foot with a pistol is a classic example.

FAST AND SLOW THINKING

Many problems occur when high-risk tasks are performed mindlessly. Mindfulness is an ongoing state of alertness, an active intuition.[15] Relative to safety, the mind must be nimble, alert to threats to assets, the occurrence of pathways during work, planned and unplanned. As mentioned previously, work must slow down for high-risk tasks when pathways have been created. Dr. Daniel Kahneman, in his landmark book, *Thinking, Fast and Slow*, characterizes human tendencies in response to

TABLE 2.2
Descriptions of Fast and Slow Thinking.

Fast Thinking (FT)	Slow Thinking (ST)
• Reflexive, usually accurate for experts in their knowledge and skill domain	• Reflective, prudent, and deliberate
• Always ON	• Turned ON and OFF (active use of working memory)
• Intuitive, pattern recognition	• Rational and logical
• Easy (skill-based performance)	• Effortful (rule- and knowledge-based performance)
• Quick (efficient)	• Protracted (thorough)
• Influenced by expertise, emotions, priming, instinct, beliefs, biases, and heuristics	• Influenced by knowledge, facts, rules, and mental models
	• Invoked by intuition, novelty, danger, and learning

Note: FT is an efficient response to the perception of risk. ST tends to be thorough in responding to an existing threat.

risk-laden decisions. Dr. Kahneman describes two modes of thinking at work in the mind, which are described more explicitly in Table 2.2[16]:

- *Fast thinking (FT)*—A subconscious approach to thought, quickly making intuitive decisions with little or no effort or deliberate attention, often automatically with no obvious awareness of voluntary control
- *Slow thinking (ST)*—A conscious, analytical approach to decision-making; mindfulness characterized by attention to details and orderly concentration on an issue—active use of working memory

Chapter 5 describes more about the application of ST using **Hu** Tools. **Hu** Tools trigger RISK-BASED THINKING, which is ST.

EXPERT INTUITION

Everyone has this "inner voice" we call intuition, often whispering things to us, sometimes called a "gut feeling," but otherwise known as instinct. Problem is often we don't listen to it until it becomes a shout, and then it's too late. Intuition is the perception of a situation without the support of conscious reasoning—thinking that is ongoing, yet unconscious. Intuition is FT. What makes intuition *expert intuition* is the depth of knowledge and experience accumulated by the individual—recognition of patterns accumulated over time. Expert intuition has been corroborated as an effective way of knowing and recognizing impending or potentially harmful situations.[17] In addition to the names listed earlier, expert intuition is also known as questioning attitude, internal risk monitor, and chronic unease. Despite what it's called, it is developed through in-depth technical education and training, mentoring, recurrent training, prework discussions, and proficiency on the job—practice—lots of it. It reinforces a mindfulness attuned to impending transfers of energy, movements of matter, or transmissions of information around key assets. Why is expert intuition so important?

Ideally, all known CRITICAL STEPS are denoted in approved work documents, but some are hidden. Despite doing their best, engineers, managers, planners, and procedure writers—given their experience, their understanding of the technology, available information, and their assumptions about the work—regularly produce guidance inconsistent with what frontline workers actually encounter. CRITICAL STEPS may be embedded in chunks of work that is considered skill-of-the-craft, which are not written out in detail. Similarly, supervisors and frontline workers could simply overlook one or more CRITICAL STEPS during their individual preparation and the team's prework discussion. Finally, work conditions may change or differ from what was assumed in the procedure or originally planned. Therefore, frontline workers can never let their guard down when working with important assets and/or hazardous processes.

Question: Can expert intuition be managed—can it be improved? It depends. Generally, the bases for intuition change slowly; some aspects are unchangeable. According to researchers, intuition is not readily educable, but it can be cultivated.[18,19] Truly, it takes years to develop technical expertise. However, frontline workers can be primed to detect threats for work in the near term. In a team environment, intuition can be augmented by conversations about the work to be accomplished during a prework discussion and during work with frequent in-field, group conversations that boost situation awareness and improve the accuracy of mental models of work in progress. Training, practice, mentoring, and experience are the best bets for the long term. The most amenable aspect of intuition in the short term is emotion. You may be able to influence a frontline worker's mindful wariness by imparting a bit of fear of what could go wrong during an upcoming work activity referring to operating experience, if available.[20]

CONVERSATIONS CREATE SAFETY

Communication is the lifeblood of high-risk operations. Communication conveys information and meaning most effectively through conversation between persons to create *shared understanding*. Like an engine's oil pump, conversation enables the flow of information that sustains successful work. Without an oil pump or if there are blockages in the system, it doesn't matter how much oil is in the engine's sump. If a person has information but doesn't share it, there is no communication. Conversations, especially face-to-face, enhance RISK-BASED THINKING about CRITICAL STEPS.

Safety is what people *do* to protect assets, and people act on their mental models—their understanding of how things work. The likelihood for success, and likewise avoiding harm, is influenced by accurate mental models. Process mental models are constructed by effective technical training. Situation awareness—mental models of current work processes and workplace conditions—is formed through a combination of technical expertise, expert intuition, and conversations. Too often, the mental models of what is happening differ among the members of a work group or even among organizational units. Only currently accurate mental models are useful. Reality is what you bump into when mental models are wrong. To remain accurate, they must be updated constantly to match the context of the technology, the work, the workplace, and the work group. This occurs with careful consideration of what must go right from conversations about what is happening and what is needed.[21] When considering the overall contribution to safety and productivity, the constant updating of a team's or work group's collective mental picture of the work and the real-time risks enhances their capacity to detect the unexpected and to form a path to success.

Robust workplace conversations foster openness, candor, and informality.[22] Though important at times, formality tends to choke the amount of information disclosed; people say only what is required. Also, "professional courtesy" can be hazardous, passing along only good news, not wanting to upset the boss, arousing unwanted attention. On the contrary, despite rank and time on the job, healthy workplace conversations should stimulate questions (and answers), new ideas, and new insights, adding depth and richness of understanding and enabling engagement. Such conversations cannot occur if people are more concerned with protecting themselves from the ire of their bosses or ridicule from co-workers. If people fear repercussions, they will not openly share what they know. Safety, quality, reliability, and even productivity will suffer.

Too often we have heard, "Be safe out there," "Pay attention," "Follow procedures," or "Don't make mistakes." Though inspirational, these words are about as effective as putting up signs around your plant that read, "Safety First!" These words and these types of signs can inhibit our mindfulness and make safety a four-letter word in the minds of workers. Instead, talking about CRITICAL STEPS and RIAs adds specificity to their understanding of safety, avoids generalities, and enhances mental engagement when done regularly. It's important from a risk perspective that conversations related to CRITICAL STEPS end with closure—who does what, when, and how. The way frontline workers talk about their work either keeps them alert to the dangers at CRITICAL STEPS or allows complacency without their realizing it.

We believe managers and executives have a moral responsibility to remove every impediment to the flow of information. The lubricating quality of oil goes bad over time. If truth and facts are replaced with generalities, half-truths, and assumptions, communication is suspect, and the organization will suffer for it. Therefore, managers must instill a "will to communicate" throughout the organization.[23] Managers must establish structures that reinforce communication, eliminate obstacles to communication, and monitor the health of communications.

Remember, *it's not who's right, it's what's right*. The nature of social interactions across group boundaries has been studied extensively.[24,25] Research supports the importance of clear, factual, and uninhibited conversations about what must go right among informed and technically competent workers and among organizational groups. Diversity of insight leads to safety and success, yet conversations take time. Interpersonal skills, diversity, and healthy relationships strengthen conversations. How to develop these is beyond the scope of the book; however, there are good texts written on these topics by Aubrey Daniels, Edgar Schein, and Rosa Antonia Carrillo, to name some popular authors.

KEY TAKEAWAYS

1. Humans are key sources of both risk (hazards) and resilience (heroes).
2. In an operational environment, human error is better thought of as a loss of control of work—of transfers of energy, movements of matter, or transmissions of information.

3. Production, risk, and safety happen at the same time. Risk exists at the convergence of asset, hazard, and human interaction.
4. A pathway is an operational situation in which an asset's transformation (change in state) is poised (exposed) to occur by either a transfer of energy, a movement of matter, or a transmission of information.
5. A touchpoint is a human interaction with an asset or hazard that changes the state of the asset through work or the control of a hazard's release.
6. Positive control of a touchpoint is most important when a pathway exists between an asset and a hazard.
7. Both the establishment of pathways and the occurrence of touchpoints are normal and necessary for the organization's success. No work would happen otherwise.
8. A CRITICAL STEP exists when there is a pathway for work dependent on one human action, a touchpoint, those human actions that involve rapid system responsiveness and severe consequences to assets when control is lost.
9. Thinking must slow down for high-risk tasks when pathways have been created.
10. Expert intuition enhances the recognition of CRITICAL STEPS.
11. Workplace conversations foster the flow of information about what's really happening, enhancing situation awareness. This is contingent on the degree of openness, candor, and informality.

CHECKS FOR UNDERSTANDING

1. While pulling the starter cord from the top of the engine of an old-style lawnmower, one foot is under the blade housing.
   a. Does a pathway for harm exist? If so, what is the asset and the hazard?
   b. Are there any touchpoints with an asset or hazard or both? If so, what are they?
2. True or False. A pathway for electrocution exists when an electrician is about to touch an exposed conductor with a test probe while performing a voltage measurement on an energized 120 vac circuit.
3. Yes or No? Walking down a long flight of stairs is a series of CRITICAL STEPS?

(See Appendix 3 for answers.)

THINGS YOU CAN DO TOMORROW

1. Relative to the organization's event analysis practices, consider reframing "human error" as a "loss of control." What might be the ramifications to systems learning if such a change was made?
2. Using the structure and headings of Table 2.1, fill out the table for explicit high-risk work activities or a recent event for specific work groups. Considering the information, discuss how the risk of losing control would be managed using the *Hu Risk Management Model*.

3. In a gathering with first-line supervisors (managers), ask them how they "manage" the risk of human error in their high-risk work tasks? Introduce the concept of CRITICAL STEPS to them, soliciting their ideas on how to apply it to their work.

4. During any production meeting, listen for whether safety is separate or part of the conversation. Do meetings start with a "safety moment," followed by the "real work?" Or is protection of assets part and parcel with talk about the production objectives? Do they understand that production and safety happen at the same time?

5. Recollect close calls from your personal life when your "inner voice" whispered to you just before the incident occurred. Ask yourself how a conscious transition to ST could have influenced the outcome.

6. Brainstorm a list of activities that can be performed with FT. With ST? Afterward, discuss why it's acceptable/unacceptable.

REFERENCES

1 Reason, J., and Hobbs, A. (2003). *Managing Maintenance Error: A Practical Guide.* Aldershot: Ashgate (p. 96).

2 U.S. General Accounting Office (1983, June). *Operation Desert Storm—Apache Helicopter Fratricide Incident, Report to the Chairman, Subcommittee on Oversight and Investigations,* Committee on Energy and Commerce, House of Representatives (GAO/OSI-93-4) (p. 46).

3 Sgobba, T. (ed. in chief), et al. (2018). *Space Safety and Human Performance.* Cambridge, MA: Butterworth-Heinemann (p. 283). The authors refer to the actualization of three basic elements as the hazard triangle: 1) hazardous element (source of harm), 2) target element (asset), and 3) initiating element (human or equipment that triggers event).

4 Viner, D. (2015). *Occupational Risk Control: Predicting and Preventing the Unwanted.* Farnham, UK: Gower (pp. 33–37, 42–44).

5 Leveson, N. (2011). *Engineering a Safer World.* Cambridge, MA: MIT Press (pp. 67, 75).

6 Howlett, H. (1995). *The Industrial Operator's Handbook: A Systematic Approach to Industrial Operations.* Pocatello: Techstar (pp. 65–67, 74–75).

7 Corcoran, W. (2016, August). 'An Inescapable of the Safe Operating Envelope (SOE).' *The Firebird Forum,* 19(8).

8 Viner, D. (2015). *Occupational Risk Control: Predicting and Preventing the Unwanted.* Farnham: Gower (p. 34).

9 Ibid. (pp. 70–72).

10 Ibid. (p. 112).

11 Kletz, T. (2001). *An Engineer's View of Human Error.* London: CRC Press (p. 145).

12 Box, G., et al. (1987). *Empirical Model-Building and Response Surfaces.* New York: Wiley (p. 424).

13 Ibid. (p. 43).

14 Perrow, C. (1999). *Normal Accidents: Living with High-Risk Technologies.* Princeton: Princeton University Press (pp. 4–6, 89–93). A tight coupling situation is characterized as one in which elements of the operations have direct interactions, which cannot be isolated, and failure with one element quickly results in failure of other elements of the system; no recovery is possible. On the contrary, loose coupling interactions respond opposite of tight coupling where recovery is possible; they are reversible.

15 Langer, E. (1989). *Mindfulness*. Boston: Addison-Wesley (pp. 115–119).
16 Kahneman, D. (2011). *Thinking, Fast and Slow*. New York: Farrar (pp. 20–24).
17 Salas, E., Rosen, M., and DeazGranados, D. (2009). 'Expertise-Based Intuition and Decision Making in Organizations.' *Journal of Management*, 36(4): 941–973. https://doi.org/10.1177/0149206309350084.
18 Kahneman, D. (2011). *Thinking, Fast and Slow*. New York: Farrar (p. 417).
19 Gladwell, M. (2005). *Blink: The Power of Thinking Without Thinking*. New York: Little, Brown and Co. (pp. 15–16).
20 Reason, J. (2016). *Organizational Accidents Revisited*. Burlington: Ashgate (p. 118).
21 McChrystal, S. (2015). *Team of Teams: New Rules of Engagement for a Complex World*. New York: Portfolio/Penguin (pp. 167–169).
22 Bossidy, L., and Charan, R. (2002). *Execution: The Discipline of Getting Things Done*. New York: Crown (pp. 102–105).
23 Allinson, R. (1993). *Global Disasters: Inquiries into Management Ethics*. Singapore: Simon & Schuster (pp. 41–43).
24 Vaughan, D. (1996). *The Challenger Launch Decision: Risky Technology, Culture, and Deviance at NASA*. Chicago: University of Chicago Press (pp. 238–277).
25 Weick, K., Sutcliffe, K. M., and Obstfeld, D. (2005). 'Organizing and the Process of Sensemaking.' *Organization Science*, 16(4): 409–421.

3 The Work Execution Process

If you can't describe what you are doing as a process, you don't know what you're doing.

**—Dr. W. Edwards Deming Author and
Consultant in Quality Management**

Spaceflight is terribly unforgiving of carelessness, incapacity, and neglect.[1]

—Gene Kranz NASA Mission Control Apollo 1 Mission, 1967

AN ENGINEERING TEST GONE WRONG

An equipment operator and a system engineer were tasked with measuring pressures at various points of a process piping system during a refinery optimization test.* The engineer collected data, while the operator manipulated system valves to connect and disconnect the test rig. They measured process pressure at various locations in the system to validate the system's capacity to operate at a higher pressure. The process fluid, a thick, wax-like product, prone to fouling instrumentation lines and valves, was operating at roughly 320 degrees Fahrenheit and 150 psig.

They had collected pressure data for 40 locations during the morning of their day shift. However, their work schedule was delayed an hour due to clogs in the test rig hoses, which had to be cleared before continuing the test. Both the operator and the engineer felt hurried because of the delay, wanting to complete the work before the end of the shift. After a brief lunch break, the operator and the engineer completed three more locations. While disconnecting the test rig hose, the test rig blew off (loss of containment), spraying high-temperature product on both individuals. Although both individuals were wearing coveralls, gloves, safety glasses, and face shields, they still received second-degree burns on their legs from the hot wax-like liquid.

At the time, the operator was using a test rig with no pressure bleed valve (see diagram in Figure 3.1). A bleed valve to control depressurizing the test

---

* Refinery optimization testing is done to increase system production without major system modifications. Optimization requires accurate measurements of flows, levels, pressures, and temperatures, throughout the system. Extremely accurate measuring devices are used instead of less accurate built-in process instrumentation.

DOI: 10.1201/9781003220213-3

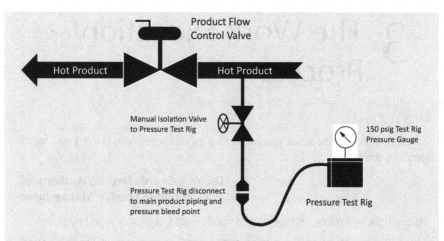

FIGURE 3.1 Simple, one-line diagram of pressure test rig connection with the process system.

rig was not installed because of repeated clogging of the test rig hoses. While thinking ahead about the next test point, the operator did not notice that pressure indicated on the test rig pressure gauge did not decrease as was the case with the previous pressure test points. As he began to remove the test rig, the operator expected that loosening the connection would release a minor amount of product trapped between the isolation valve and the test rig and bleed off any trapped pressure. Normally, the local manual isolation valve, between the process system and the test rig, was closed before the test rig was connected and disconnected between tests. This time the valve was full open; the test rig was still lined up to the process under full system pressure. He forgot to close it.

In all 43 previous tests, the operator connected and disconnected the test rig successfully. Each time, the operator manually closed the isolation valve before disconnecting the test rig. It was within an arm's length, and the operator could see the valve while removing the test rig. Yet the operator was unaware of the isolation valve's position. If the operator had known the condition of this isolation valve, it is unlikely he would have disconnected the test rig prematurely.

THE THREE PHASES OF WORK

The *Work Execution Process* focuses on the "sharp end"—that part of the organization's system where frontline personnel perform work. This is where the organization's transformation (value creation) processes occur. To enhance the ability to manage human performance for high-risk work systematically, you need a clear picture of *work as a process*. A process is a repeatable sequence of functions or actions ordered to achieve a desired outcome safely and reliably. There are three phases of work at the sharp end: 1) preparation, 2) execution, and 3) learning (see Figure 3.2). In principle, the process of accomplishing work is the same for every task, yet the details of work will differ significantly between industries and technologies.

When considered separately, the three phases of the work execution process promote greater consistency in managing human performance risk, which naturally accommodates RISK-BASED THINKING. Each phase of work offers opportunities to *anticipate* what to expect, to *monitor* significant changes to assets and hazards, to *respond* in ways to retain positive control of human activities and protect assets from harm, and to *learn* from the past and present to improve the safety, reliability, and productivity of future performance. Depending on the industry or technology, certain practices/tools can be overlayed each phase of work to guide the identification and control of CRITICAL STEPS. For a time-proven application of the *Work Execution Process* as a structure for managing safety, see the DOE's *Integrated Safety Management System*.[2] The following sections describe the phase-specific activities that enhance safety, reliability, quality, and even productivity.

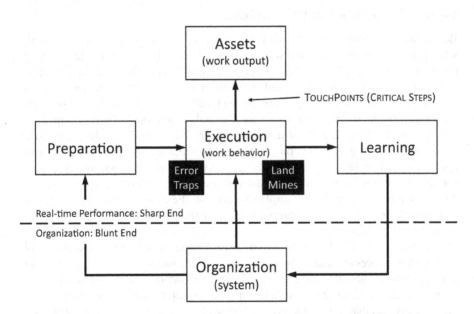

FIGURE 3.2 The *Work Execution Process*. Work (transformation) occurs at the sharp end of the organization where frontline workers "touch" things (human actions). Resilient organizations allot time to preparation and learning as well as to production work.

PREPARATION

Time is allotted to workers and their supervisors for preparation for high-risk work, to get ready mentally for the work at hand. Experience has shown that mental preparation improves the likelihood that people will succeed with an assigned task.[3] This is a widespread practice in competitive sports. Before engaging in hands-on work, workers take time to carefully review procedures and other relevant guidance to develop a clear mental picture of the planned work, current system conditions and equipment status, desired work outputs, and the plan on how to achieve those. If performing shift work, turnover from the off-going shift is also reviewed and questioned. The work-related information reviewed gives each person a clear understanding of what is to be accomplished and what is to be avoided. Preparation involves reviewing and understanding the following[4]:

- Objectives of work (accomplishments and criteria for success)
- Anticipated initial and final conditions
- Precautions and limitations (constraints)
- Team member roles and responsibilities
- Sequence of work tasks and relevant procedures
- Tools and resources required
- Risks and hazards (emergency preparedness)

WHAT TO ACCOMPLISH AND WHAT TO AVOID

Before starting work, frontline workers should become intimately familiar with what is to be *accomplished* and what is to be *avoided*. In addition to the work at hand, frontline workers should appreciate the business purpose of their work—the value they intend to create and the related success criteria. Knowing what is to be accomplished, top performers generally have a better sense of what needs to go right and more readily notice signals when things begin to unravel. In their book *Managing the Unexpected*, Karl Weick and Kathleen Sutcliffe make the following observation of HROs: "A reliable system is one that can spot an action or function *going* wrong, not an action gone wrong." Top performers in HROs train their "internal risk monitor" to detect the presence of danger. Knowing specific accomplishments and the assets' critical parameters help workers detect when an activity is going wrong—steering off course—and avoid failure proactively.

Similarly, workers and their supervisors must consider what to avoid; loss of control of hazardous processes, and harm to key assets. Such foreknowledge pinpoints what must go right and how to exercise positive control of built-in hazards that will be used to create value. However, it's not uncommon workers skip the part about failure or else give it lip service, that is, to *fail safely*.[5] Failing safely is the capacity to limit harm done, assuming there is a loss of control or equipment failure, to place an asset in a safe state before the onset of serious harm, to keep something bad that has happened from getting worse. This includes the capacity to recover from either a loss of control or the harm realized. Dr. James Reason also suggested providing means of escape and rescue.[6] More on failing safely in a few paragraphs. But, before starting work, workers would do well to ask themselves what could be done in advance to preserve positive control of hazardous work processes and to limit the harm done should they lose control. Both what to accomplish and what to avoid are important to know.

THE PREWORK DISCUSSION

An important success factor for work teams involves a robust dialogue that explores the reality of the work, the workplace, and the workers' readiness. A prework discussion is a meeting, sometimes called a huddle, prejob briefing, tailboard, or tailgate. It is a conversation about the work at hand between the assigned worker, or work group, and the responsible supervisor. The meeting should take place in a quiet location where it will not be interrupted. During this meeting, supervisors and frontline workers explicitly review what is to be accomplished and what is to be avoided.

In most cases, prework discussions occur on the day of the job and usually last 15–30 minutes—depending on the work's risk and complexity. Occasionally, it may be prudent to conduct a physical walkdown of the task in the workplace. It should be obvious that a prework discussion is not the time and place to conduct in-depth training, plan, or study procedures. These preparations are done before the briefing, during work planning. The prework discussion validates preceding work planning activities. Note that the time devoted to a prework discussion is part of the job, not separate from the job, and should be incorporated into the duration of a job when scheduling work! (The same can be said for postwork reviews, which are discussed later.)

A prework discussion involves all four aspects of RISK-BASED THINKING: anticipate, monitor, respond, and learn. Learning from events similar to the work at hand provides important insights into the risks involved. It is important for the workers to consider explicitly how their actions will affect key assets—what to expect and what to pay attention to. Among other work-related responsibilities, the prework discussion helps field workers know in advance what to do to maintain positive control of CRITICAL STEPS and how to protect assets if things don't go right. Prework discussions enhance the worker's capacity to adapt when necessary. **RU-SAFE**, an **Hu** Tool adapted from methods used in other industries, incorporates the preceding aspects of RISK-BASED THINKING into a structured approach to guide the conversation during a preview of a high-risk work activity*:

1. _Recognize_ assets important to safety, reliability, quality, production, etc., and their relevant critical parameters.
2. _Understand_ workplace hazards and built-in operational hazards to each asset, considering relevant lessons learned from previous jobs and events.

---

* Readers are given permission to use this **Hu** Tool as written on the condition that attribution is given to this book and its authors.

3. *Summarize* the CRITICAL STEPS and related preconditions (risk-important conditions) necessary for safety, established earlier by RIAs, and the means of positive control for each CRITICAL STEP.
4. *Anticipate* ways of losing control for each CRITICAL STEP, highlighting menacing error traps.
5. *Foresee* worst-case consequences for each asset, upon losing control of a CRITICAL STEP.
6. *Evaluate* the controls, barriers, and safeguards at each CRITICAL STEP, including contingencies, STOP-work criteria, communication methods, and emergency protocols to fail safely.

Individuals can preview their work using the RU-SAFE checklist not only during work preparation, but also during the work planning process, the procedure development process, and even during work in the field. It can help anyone think through one's actions before performing them—anytime, anywhere. However, the effectiveness of RU-SAFE depends greatly on the user's technical expertise, without which the worker may not recognize the limits of safety for an asset—the edges of the cliff defined by its critical parameters. When the prework discussion is complete, there is no confusion about:

- Work to be accomplished; expected work outputs or end states, and success criteria
- Assets, their key critical parameters to pay attention to during the work, and their respective hazards (pathways)
- CRITICAL STEPS (touchpoints) and related RIAs
- Prevalent error traps and possible ways of losing control at each CRITICAL STEP
- Means of positive control of hazards and CRITICAL STEPS (e.g., **Hu** Tools)
- Lessons learned from previous experience relevant to the anticipated CRITICAL STEPS
- STOP-work criteria for each CRITICAL STEP
- Contingencies for a loss of control at each CRITICAL STEP and means to fail safely
- Means and locations of emergency communication devices and who to ask for help when questions arise

PREPARE TO FAIL SAFELY—ANSWERING THE RIGHT QUESTIONS

In engineering, a fail-safe is a *design feature* or *practice* in which, in an event, the system inherently responds in a way that will cause minimal or no harm. This is different from "inherently safe," which means *no* harm occurs. Failing safely does not mean that harmful consequences are impossible or improbable, but that the system's design mitigates harm as an outcome of the system's failure.[7] Similar to a safety net below a high-wire circus act, to fail safely means to place an asset and its associated systems in a safe mode after a loss of control or equipment fault but before significant onset of harm, if possible. An engineered example is an electrical ground fault circuit interrupter (GFCI). A GFCI is a design feature of wall sockets and switches that protects human beings from electrical shocks by tripping the power supply when a ground is detected at the user end. The onset of harm after losing control at a CRITICAL STEP is not preventable—occasionally, there are tasks that have no fail-safe (no harm) alternative.

Frontline workers are tempted to dispense with conversations about what to avoid when they perceive the work to be simple or routine.[8] To prompt this conversation, a question that is seldom, if ever, asked in a prework discussion is *"What must go absolutely right, the first time and every time?"* This question evokes an important set of answers; it helps identify the most important steps or actions for a given task—CRITICAL STEPS and the positive actions required to make sure things go right—who does what, to what, with what, how, and when. What if the equipment operator and system engineer had asked this question before starting their pressure test?

To improve the chances of failing safely, workers should think about and discuss answers to questions, such as *"What's the worst that can happen?"* *"What if . . . ?"* and *"What could go wrong, and when it does, what will you do?"* Questions such as these attempt to foresee consequences of a loss of control. Answers can be refined by posing these same questions differently, focusing on assets that need protection. What is the worst thing that can happen, when [*describe task*] is done? What will you do when [*describe occurrence*]? Each person should be asked these questions and expected to answer them, especially the last question—no exceptions. Their verbal answers instill a mindfulness (intuition) about what to pay attention to and what to do. By having the right mindset, workers are more likely to respond faster and more accurately, limiting the degree of harm or damage to an asset.

We authors have observed hundreds of jobs and their prework discussions. Too often, when workers are asked, "What's the worst thing that can happen?" it's not uncommon to hear, "Someone could get hurt." The real problem with such an answer is that there is no follow-up conversation on what to do to avoid the outcome! Mindless or halfhearted answers to this question are offered because the question is asked in an insincere, perfunctory manner simply to complete the paperwork. Consequently, no real conversation or learning takes place. The following real-life conversation illustrates the problem that so often plagues the effectiveness of prework discussions.

ELECTRICIANS UNPREPARED TO FAIL SAFELY

Recently, I (Ron) observed a prework discussion at a major electric utility that involved several electricians changing the turbine-generator exciter* brushes while the generator was still rotating and producing power. The meeting itself covered most of the key elements defined by the company's

---

* An electric generator must have relative motion between a magnetic field and a coil of wire to produce an output voltage. The magnetic field is produced either by a permanent magnet or small generator. Power plant generators use a small direct current generator attached to the generator's rotor to provide the electrical current to send through this coil that produces the magnetic field in the main generator rather than a permanent magnet. This small generator is called an exciter generator.

policy for a prework discussion. However, the last question on the checklist was not asked: "What is the worst thing that can happen?" When prompted, the supervisor asked the question; and as expected, the workers gave the patent answer, "Someone could get killed." I interjected and posed the same basic question but with a couple of caveats, "What's the worst thing that can happen given the hazards that exist? What are they?" The workers accurately described two primary sources of energy: first, the electrical hazard from the exciter itself; and second, the mechanical hazard from the rotating shaft (1,800 rpm). These experienced workers already knew that electrical shock or electrocution and physical harm from the rotating equipment were obvious hazards but did not call them out explicitly. "It's common sense, right?"

The first step to preparing to fail safely is to create awareness and knowledge. The three workers were asked what they would *do* if the senior electrician was shocked or electrocuted. The designated tool runner, who was to be outside the work boundary, was asked, "What would you do if this happened?" He replied, "I'd call 911. No, wait—it is 9–911." The supervisor corrected the runner with, "You call 911 because it will reach our local emergency response team; 9–911 would call the local dispatch in town." When the designated safety watch, who was to be inside the safe work boundary, was asked how he would respond, he replied hesitantly with, "I just qualified CPR (cardiopulmonary resuscitation), I can't remember . . ., I think . . . I might . . .;" and on it went until I politely interrupted him. I asked the safety watch to calm down and think about the proper response based on his training in CPR. After a long pause, the individual responded properly. When I asked the senior technician if he was comfortable, he responded, "Hell no, not at all."

I then asked the workers, "Let's consider the rotating equipment source of energy. The senior technician gets a limb torn off his body. What will you do?" The runner identified quickly and correctly, "I'd call 911." Nervous, the safety watch responded the same way as for the first question, with concern, not knowing the correct response. Again, I was able to calm him down and take him through this scenario, and he was able to offer a proper answer. This time when the senior technician was asked if he was comfortable, his response was, "I wasn't at first, but now I am."

The preceding account illustrates the importance of robust conversations.[9] If either scenario had occurred, a loss of life was a real possibility; these electricians were not mentally prepared to respond properly and quickly. Without the appropriate mindset, a significant delay in response would have dramatically reduced the chance of survival had an injury occurred in either scenario. If Ron had not intervened, this shortcoming would have prevailed in future prework discussions. Research suggests that we can train our brains to react to a sudden upset properly by mentally rehearsing an emergency response, rather than letting our instinctive survival response dictate our actions.[10] Managers and supervisors must continuously challenge their workers during work preparation to be ready to fail safely and to continuously urge one another to practice RISK-BASED THINKING (anticipate, monitor, respond, and learn) during every phase of work.

Let's turn our attention back to the refinery optimization test (burn event) described at the beginning of the chapter. We are not sure the operator and engineer took the time to conduct a prework discussion as part of their preparation, but let's infer some observations about what probably happened if they had one.

- We know the test rig did not have an installed pressure bleed valve, requiring the operator to manually bleed trapped product when disconnecting the test rig hose. Did they talk about the risk of losing containment when disconnecting the test rig from the system?
- Did the operator and engineer talk about their task from a risk perspective before starting work? Apparently, their conversation did not explore the means to safely control the hot wax-like fluid given that a vent was not feasible and the system's history of clogging.
- Did they review previous events relevant to their task? If so, they could have made plans on what each person could do to avoid similar mistakes or outcomes.
- Presumably, the test rig hoses were ready for the tests and clear of obstructions. However, they had no spare hoses, since the one hose they were using clogged at one point and had to be cleared of wax before they could resume the task.
- It's likely the prework discussion did not explicitly identify the *most important action* that had to go right to avoid a loss of containment, disconnecting the test rig (the CRITICAL STEP).
- In light of the preceding item, they most likely did not discuss the importance of shutting the isolation valve *before* disconnecting the test rig (the related RIA).
- They did not assign responsibilities explicitly—who would do what and when—for safe execution of their work.

EXECUTION

In his book *Flawless Execution*, James Murphy, a former F-15 fighter pilot, suggests that execution is nothing more than *"flying the briefing."*[11] Fighter pilots execute their missions based on the briefing—*work-as-imagined* (planned). It is the intent of frontline workers and their supervisors to perform the *work-as-planned*. But, as the Colonel (in Chapter 1) realized, nothing is always as it seems—*work-as-done* is rarely the same as *work-as-imagined*. War plans rarely survive first contact with the enemy.

The word "execution" has special meaning in the workplace. It is not to be confused with the infliction of capital punishment or the process of carrying out a judgment or sentence of a court of law. In the context of work, "execution" refers to the performance of planned work to accomplish a business goal—to create value.

After their preparation, workers should know what is to be accomplished, what is to be avoided, and what to monitor—what to pay attention to—as they begin their tasks. They have considered how they will respond to maintain positive control of CRITICAL STEPS and how to protect assets should they lose control.

When arriving at the job site, workers inspect the workplace to confirm as-found conditions correspond with anticipated initial conditions. Time is taken to identify workplace hazards not directly involved with the work at hand as well as work-related, built-in hazards. They acknowledge the assets and hazards they expect to work with, noting assets and hazards not reviewed during the prework discussion. Workers should be careful to avoid rationalizing what they see and hear; if anything is unusual or confusing, STOP and get help. This practice—a timeout is commonly referred to as a "two-minute drill," "take two," or "take a minute" to mentally inspect the workplace before starting work.

When real work is performed, frontline workers come into close physical contact with the organization's assets, products, services, and built-in operational hazards; this is no place or time to make assumptions. Knowing that things are not always as they seem, workers stay alert for surprise pathways (landmines). They start touching and altering things guided by specified technical processes that involve various transfers of energy, movements of matter, and/or transmissions of information. As work progresses, things change. Several non-technical methods, such as **Hu** Tools, are mindfully applied to exercise positive control at RIAs and their CRITICAL STEPS. This is where losing control can be most detrimental to safety, quality, reliability, production, and even profitability. Hopefully, there are no surprises. If there are, they adapt to their circumstances to protect assets from harm.

A MULE'S TALE

Handling horses and mules always has the potential for harm. These animals are powerful, and things can go bad fast. Gary, my brother, called me one day and asked if I (Ron) would come over and help him get a piece of barbed wire out of his mule's tail (the hair of the skirt). Unfortunately, I said yes. We had a brief discussion of who was going to do what and when. I was going to be the person at the dangerous end of this job with scissors in my hand. My brother was going to use a twitch, which is a two-foot wooden handle with a short chain anchored to the handle to form a small loop. The chain twitch forms a loop that is twisted around the fleshy part of the animal's nose and keeps its head from moving.

My brother said he had never used one before, and he was not sure about using a twitch (my first clue). The twitch works in three ways. First, it provides some minor physical restraint. Second, it generates some discomfort for the animal, especially when first applied. This discomfort usually distracts the animal from minor manipulation or discomfort elsewhere. And third, after some time in place, the twitch causes release of endorphins (naturally analgesic painkillers).

After a brief explanation, I applied the twitch, handed the twitch to him, and said, "Now whatever you do, don't let go!" He seemed nervous and his hands were shaky (my second clue). As I went to work cutting the mule's tail to extract the barbed wire, my brother started having trouble with the twitch, and he said, "I can't keep it on" (my third clue). I yelled at him to keep it on. (By the way, not the best way to get top performance.)

Shortly thereafter, Gary told me again that he was losing his hold, the twitch slipped off the mule's nose. In the blink of an eye, the mule swung his backside around and landed a hard kick to my left thigh, dropping me to the ground. When I looked up, a second kick nicked my left ear. I scrambled to safety. After a lengthy recovery and choice words, we switched jobs and completed the task without further incident. My leg took a month to heal from a deep tissue bruise.

Ron's experience illustrates the pitfalls that can ensue after inadequate preparation. While working on the mule's tail, Gary warned Ron that he was losing control of the twitch. Ron, focused on achieving his goal, did not back away from the mule—stop the work—to help Gary restore proper control of the mule. Ron remained in the line of fire of the mule's powerful hind legs. Perhaps, Ron will consider wearing a hockey helmet next time, in the spirit of failing safely.

Recognizing an impending CRITICAL STEP offers the performer the opportunity to pause before proceeding with the action, slowing down the activity and shifting from fast, automatic thinking to slow, deliberate thinking, to verify that safe and proper conditions exist before proceeding with the work. Preferably, each CRITICAL STEP is verbally announced, bringing it to the attention of team members in some manner ("Step 'X' has been reached, a hold-point, CRITICAL STEP") so everyone on the team knows a crucial point in the job has been reached. To prove it is safe to proceed with a CRITICAL STEP, proper *conditions* established by preceding RIAs are verified.

At the beginning of this chapter, the lead-in story involved second-degree burns suffered by an operator and an engineer. Let's review the execution phase of their work—collecting pressure data on a system running at operational temperature and pressure.

- Forty-three tests had been performed properly, including 43 CRITICAL STEPS. Human beings are susceptible to distraction and inattention for repetitive tasks—complacency from routine.
- The clogged test rig hose delayed their effort for an hour because the hose had to be cleared before the task was resumed. This imposed time pressure on the individuals, urging them to hurry—an error trap.
- The operator did not habitually pause to confirm the pressure in the test rig before disconnecting it from the system—not recognized as a CRITICAL STEP.

- What method(s) did the individuals use to avoid a leak (at 150 psig), a loss of containment? Did they consider **Hu** Tools, various means of exercising RISK-BASED THINKING? Self-checking? Peer-checking? Callouts? Three-part communication?
- The engineer did not regularly check that the isolation valve closed before the operator disconnected the test rig—no peer-check.
- The operator performed 44 CRITICAL STEPS. The isolation valve was inadvertently left open on the last test.

The operator could have called out the CRITICAL STEP, "Disconnecting the test rig," which may have alerted the engineer of the operator's intended actions. Perhaps, a verbal callout would have triggered their *intuition* about the situation to think about the status of the isolation valve before disconnecting the test rig hose. The role of **Hu** Tools used during work execution alert workers to slow down their thinking (see Chapter 5).

LEARNING

When work is completed, it is always tempting to quickly move on to the next task in a market-driven production environment. The learning phase is often overlooked or omitted for the sake of productivity. If a job is important enough to devote time to a prework discussion, then it is important enough to give time to a postwork review in the *Learning phase* (knowing what to change). If the postwork review is neglected, managers will miss an important opportunity to know more about the effectiveness of their systems.

Learning is a feature of RISK-BASED THINKING that involves *knowing* 1) what has happened, 2) what is happening, and 3) what to change going forward.[12] Learning occurs in every phase and is particularly strengthened with ongoing, robust conversations. Learning from previous events (lessons learned or operating experience) and the personal experience of others is an expectation during prework discussions in the *Preparation* phase (knowing what has happened). During the *Execution* phase, through situation awareness and frequent conversations with co-workers, workers preserve an accurate understanding of evolving technical processes and current conditions of assets and their hazards. When managers and supervisors are present, workers receive individualized feedback about their performance during field observations. Managers, in turn, learn about the functioning of their various management systems (knowing what is happening).

Fundamentally, learning depends on the occurrence of and response to *feedback*. Without feedback, there is no opportunity for a change in behavior, whether for the system or for the individual. Learning is more effective when it involves *two-way* feedback: feedback from workers to managers and feedback from managers to workers. Both are necessary. Work-related feedback 1) provides managers with opportunities to improve organizational systems among other systemic factors and 2) presents occasions to reinforce superior performance and to coach individual worker performance involved in a work activity. Although it is beyond the scope of this book to discuss in detail, the occurrence and quality of feedback depend significantly on the climate in which people are willing and able to talk about what is going on, without fear—respectful, candid conversations. A safe climate—sometimes referred to as a *just culture*—grows from adherence to Principle 1 (see Appendix 2).* Presuming people are treated with dignity and respect, there are three important means (tools) for creating opportunities for feedback:

- *Timeout*—This is a brief work stoppage to review what must go right and related preconditions, to warn others of changing workplace conditions, to clarify responsibilities, etc., all learning oriented.[13] Such conversations help people to update their situation awareness and mental models, provide opportunities to express concerns, challenge and update assumptions, suggest performance tips, etc., with others in the work group. The intent is to create a shared understanding of the task, work environment, its risks, and individual issues. Timeouts can be scheduled to occur after a given period of time, during interruptions of work when there is a pause in the work activity, and at CRITICAL STEPS. Timeouts can be called any time, but some should be planned to occur at key decision points and always prior to performing CRITICAL STEPS (see *Hold Points* in Chapter 5).
- *Field Observation*—This activity gives managers the opportunity to acquire firsthand information about the effectiveness of work preparation, to acknowledge what is going well, and to better understand worker challenges, concerns, readiness, and actual performance. Field observations also provide insights into differences between *work-as-imagined* and *work-as-done* (see subsection "Field Observation and Feedback" in Chapter 6).
- *Postwork Review and Reporting*—Workers give feedback through structured means that provide managers with a rich, valid, and fresh source of information about task-specific conditions, procedures, resources, coordination, incentives, and disincentives and their related management systems (see the subsection "Postwork Review and Reporting" in Chapter 6). The primary aim of these postwork conversations is to identify significant and persistent differences between *work-as-done* and *work-as-imagined*, in the form of a question, "What was different between work preparation (what we planned to do) and work execution (what we actually did)?" It's also beneficial to note/document significant as-left conditions of equipment (related to assets and hazards) or the workplace to pass on to the next shift, operations, engineering, work planners, or other concerned groups. Obviously, such information must be reported (reliably communicated) to line managers to give them the opportunity to make appropriate system changes (realignments) in support of future work.

For these learning opportunities to occur regularly and effectively, the organization and its management systems must be aligned to support them. Work schedules and resources must accommodate not only prework discussions but also time for managers to observe work and to provide frontline workers the time and place to

---

* See Dr. David Marx's book, *Whack a Mole*, for practical understanding of what a "just culture" looks like and how to establish it. See also, Sidney Dekker's *Just Culture* (2nd ed.).

conduct postwork reviews. Accountability at the line manager level is necessary to ensure learning occurs for frontline workers and for the system. Follow-through on learning is ascertained through a formal corrective action/preventive action (CA/PA) process. (For more details on CA/PA, see Chapter 6.)

What opportunities for learning existed for the equipment operator and the engineer? For the organization? Obviously, there was the subsequent event investigation for the burns the individuals received—an investigation is learning late. But had there not been an event, what opportunities for learning were available before, during, and after the work? Here are some learning opportunities that the duo could have taken advantage of.

- What was different between the 43rd and 44th tests?
- What error traps existed on the last test that did not exist previously?
- Why did the pressure measurements have to be completed by end of shift?
- What was done on the 44th test that was not done on the previous tests?
- Why didn't they have spare hoses? Clogged hoses were not a new problem.
- When the test rig hoses clogged, did the operator and the engineer take advantage of the break in their routine (timeout) to talk about the risk with disconnecting the test rig hose?
- Why didn't the engineer peer-check the operator?
- Where was supervision throughout the day? No one visited the entire day to check on performance or to see if they had any needs.
- Their coveralls did not protect their legs from severe burns when sprayed with the 350-degree wax-like fluid. Why hadn't this been considered before?
- What can be done differently the next time a similar task is performed?

Assuming no event occurred and, instead, the work duo participated in a postwork discussion, the answers to the questions could have been reported to management— an opportunity for SYSTEMS LEARNING.

TABLE 3.1
Summary of the *Work Execution Process*—the Purposes, Means, and Success Criteria for Each Phase of Work.

Summary: Work Execution Process

	Preparation	Execution	Learning
Purpose	• To know what to accomplish (including criteria for success). • To know what to avoid (loss of control of hazards and harm to assets)	• To exercise positive control of known CRITICAL STEPS • To detect and resolve landmines (unknown CRITICAL STEPS) • To fail safely after a loss of control	• To identify significant differences between *Work-as-Done* (execution) and *Work-as-Imagined* (preparation) • To strengthen anticipation, monitoring, and responding • To provide feedback to line managers and to workers

Summary: Work Execution Process

	Preparation	Execution	Learning
Means	• Prework discussion (Chapter 3) • Review of work guidance and procedures • Work location walkdown	• **Hu** Tools (Chapter 5) • Chronic unease (Glossary and Appendix 2) • Hold points at CRITICAL STEPS (Chapter 5) • Positive control (Chapter 5) • Conversations (Chapter 2) • Field observations and feedback (Chapter 6) • Adaptive capacity (Chapter 6) • Expert intuition (Chapter 2)	• Timeouts (Chapter 3) • Postwork reviews (Chapter 6) • Observation and feedback (Chapter 6) • Reporting (Chapter 6)
Success criteria	• Frontline workers and their supervisors know what to expect and what to pay attention to • They know what must go right and what to do if it does not • They know the key assets to protect and their safety-critical parameters • They know how to exercise positive control of hazards • They know STOP-work criteria and contingencies to fail safely • Adequate time and place provided for prework discussion meeting	• Precision execution—positive control of CRITICAL STEPS • Preconditions of safety established for every CRITICAL STEP (Chapter 4) • Minimal harm to assets if a loss of control occurs • Mindful consideration of and response to error traps (Chapter 5) • Recognition and control of landmines (Chapter 5)	• Mental break between work completion and start of postwork discussion meeting • Sufficient time and place provided for postwork discussion meeting • System realignment that supports identification and positive control of CRITICAL STEPS, protects assets, and enhances/preserves adaptive capacity (Chapter 6) • Prompt implementation of corrective/preventive actions

Note: See referenced chapter for more details on the respective element.

KEY TAKEAWAYS

1. The *Work Execution Process* provides a mental model of human activities in the workplace, involving three general phases of work: 1) preparation, 2) execution, and 3) learning.
2. Prework discussions clearly define what is to be *accomplished* and what is to be *avoided*. Frontline workers must have a sharp vision of what success looks like for key assets involved in the work.
3. During discussions on what to avoid, everyone should be asked, *"What could go wrong, and when it does, what will you do?"*

4. The *Preparation* phase of work provides time to verify that assigned field workers know CRITICAL STEPS, how to ascertain safe preconditions exist for each, what to do to exercise positive control of their performance, and how to fail safely. Preparation primes frontline workers to expect and detect landmines (surprise CRITICAL STEPS).

5. To fail safely means to lessen the onset of harm to an asset after losing control at a CRITICAL STEP. Frontline personnel must be prepared mentally to act promptly after a loss of control at a CRITICAL STEP.

6. *Execution* is the point when real work is performed, when people come into close, physical contact (touchpoints) with the assets, products, services, and related operational hazards, things change for good or for bad. This is where losses of control (including human errors) can be most detrimental to safety, quality, reliability, production, and even profitability.

7. *Learning* has not occurred until there is a subsequent change in behavior. Learning occurs in both individuals and systems, but without feedback there is no opportunity to change. Learning occurs in all three phases.

8. If a job is important enough to have a prework discussion, it is important enough to have a postwork review.

9. Work schedules must accommodate the time needed to prepare for work, to exercise safety and reliability, to perform CRITICAL STEPS with precision, and to learn from the work done.

CHECKS FOR UNDERSTANDING

1. CRITICAL STEPS occur during which phase of the Work Execution Process?
   a. Preparation
   b. Execution
   c. Learning

2. True or False. Learning occurred when the work team recognized a problem with the job just completed and verbally reported it to their supervisor.

3. Which elements of RISK-BASED THINKING do frontline workers use during their preparation for a task? (There may be more than one answer.)
   a. Anticipate
   b. Monitor
   c. Respond
   d. Learn

(See Appendix 3 for answers.)

THINGS YOU CAN DO TOMORROW

1. Prepare a poster with the **RU-SAFE** Prework Discussion guidance. Post it on several walls of rooms used for prework discussions.

2. Post the *Work Execution Process* in workshops or anyplace work planning and preparation occur. Encourage work groups to improve their preparation for the CRITICAL STEPS of the work and to learn more deliberately from their experiences on the job.

3. Review the work planning process to understand how high-risk activities are identified, controlled, and denoted in procedures and other work documents. Determine if CRITICAL STEPS are identified and denoted. If not, why not?
4. Watch a prework discussion for a high-risk task. Assess how the work group addresses what must go right and what to avoid during the work. Do they talk about past successes and mistakes and what to do to avoid them? What about the worst that can happen given the hazards and assets for the task? Do they discuss what to do to mitigate the harm?
5. Identify how workers provide feedback on completed work. Is time allowed to discuss differences between execution, what they did, and preparation, what they planned to do?
6. Consider a task that the group stopped. Discuss what practices were responsible for helping recognize that things were beginning to unravel?
7. For the work stopped, describe how learning influenced the decision to stop the work. Share this information with the supervisor and work planners.

REFERENCES

1 Kranz, G. (2000). *Failure is Not an Option*. New York: Simon and Schuster Paperbacks (p. 204).
2 U.S. Department of Energy (2012, September). *Brochure, A Basic Overview of the Integrated Safety Management (ISM)*. Office of Environment, Health, Safety and Security. Retrieved from: www.energy.gov/ehss/downloads/brochure-basic-overview-integrated-safety-management-ism.
3 Spear, S. (2009). *The High-Velocity Edge*. New York: McGraw-Hill (p. 23).
4 Howlett, H. (1995). *The Industrial Operator's Handbook: A Systematic Approach to Industrial Operations*. Pocatello: Techstar (pp. 111–114).
5 Conklin, T. (2017). *Workplace Fatalities: Failure to Predict*. Santa Fe: PreAccident Media (p. 16).
6 Reason, J. (1997). *Managing the Risks of Organizational Accidents*. Brookfield: Ashgate (p. 7).
7 Rutherford, D., Jr. (1990, June 2). 'What Do You Mean—It's Fail Safe?' Paper presented at *1990 Rapid Transit Conference*. Hosted by American Public Transit Association. Vancouver, British Columbia, Canada.
8 Howlett, H. (1995). *The Industrial Operator's Handbook: A Systematic Approach to Industrial Operations*. Pocatello, ID: Techstar (pp. 111–114).
9 Jonassen, J., and Hollnagel, E. (2019, March). 'License to Intervene: The Role of Team Adaptation in Balancing Structure and Flexibility in Offshore Operations.' *Journal of Maritime Affairs*, 18, 103–128. https://doi.org/10.1007/s13437-019-00166-y.
10 Robinson, S. (2012). 'When Disaster Strikes: Human Behaviour in Emergency Situations.' *Journal of the Institute of Civil Protection and Emergency Management*. Retrieved from: http://clok.uclan.ac.uk/9573/. When an unexpected threat arises, usually there is limited time to decide what to do. Our natural threat response is not always what the movies depict—only a small fraction (10–15 percent) of humans respond with hysterics, such as running and screaming. Instead, most individuals freeze, taking no action. Research into human responses to surprise threats indicates that a low percentage of people (10–25 percent) respond to an emergency properly, whereas most (65–80 percent) disaster victims freeze.

11 Murphy, J. (2005). *Flawless Execution: Use the Techniques and Systems of America's Fighter Pilots to Perform at Your Peak and Win the Battles of the Business World.* New York: Regan Books (p. 19).

12 Hollnagel, E., and Woods, D. (2006). 'Epilogue: Resilience Engineering Precepts.' In: Hollnagel, E., Woods, D., and Leveson, N. (eds.). *Resilience Engineering: Concepts and Precepts.* Burlington: Ashgate (p. 348).

13 Jonassen, J., and Hollnagel, E. (2019, March). 'License to Intervene: The Role of Team Adaptation in Balancing Structure and Flexibility in Offshore Operations.' *Journal of Maritime Affairs*, 18, 103–128. https://doi.org/10.1007/s13437-019-00166-y.

4 Risk-Important Actions

It's not enough that we do our best. Sometimes we must do what is required.[1]

—Sir Winston Churchill

Speeding fines are double when workers are present.

—Road sign along Alabama highways when approaching construction zones

TIGHT GRIP NEAR SLIP[2]

While on holiday, a Tourist and his spouse went to the Alps in western Europe, and he elected to go on a hang-gliding excursion. Unfortunately, he ended up hanging on for dear life after the pilot forgot to strap him in.

While the Tourist and the pilot were still on the ground, there was no sign of trouble. However, trouble became apparent as soon as their feet left the ground—the Tourist immediately realized that his harness was not attached to the glider. The pilot forgot to secure him to the glider, which meant that his harness served no safety purpose as he hung in the air, holding on with just his hands, now hundreds of feet above the ground. The pilot realized what was happening and attempted to land the glider before going too far. Too late. He was having trouble controlling it, taking him more than two minutes until he could land. All that time, the Tourist was holding on to the landing gear with his left hand, and his right was grasping the pilot's harness. The pilot was holding him with one hand and steering the glider with the other.

Fatigued, the Tourist couldn't hold on any longer when they approached the ground. He let go of the glider when about five feet from the ground. He hit the ground at 45 mph and broke a wrist, which needed surgery afterward. The Tourist also suffered a torn tendon in the bicep of the same arm. To the Tourist's credit, he was very gracious about pilot's oversight and his resulting injuries.

Identifying CRITICAL STEPS and their related RIAs allows frontline workers to establish the preconditions for safety as well as the preconditions for work. In the Tourist's case, the CRITICAL STEP was becoming airborne. The associated RIA was attaching his harness to the glider. Frontline personnel can *prove* a task is safe by verifying the preconditions are established by the RIAs relevant to the subsequent CRITICAL STEP.

DOI: 10.1201/9781003220213-4

WHAT IS A RISK-IMPORTANT ACTION?

Fundamentally, RIAs serve two functions: create the *conditions for work* and establish the *conditions for safety*, Risk-Important Conditions. In any operational process, pathways for work are established before performing work so that energy, matter, or information are *poised* for immediate productive use. Usually, these conditions are established by one or more preceding actions as directed by the guiding procedure. Occasionally, RIAs are an element of skill-of-the-craft, which normally is not explicitly called out in a written procedure. But before performing work, people must prove it is safe to proceed by establishing or verifying the presence of proper defenses that will protect the asset by moderating the work to not exceed one or more of the asset's critical parameters. An RIA includes any one or more of the following human actions that:

- Create or remove pathways for the transfer of energy, movement of matter, or transmission of information that expose an asset to a hazard (necessary for the conduct of work)
- Increase or decrease the number of actions needed to begin work (likewise, margin for error)
- Influence the presence or effectiveness of barriers or safeguards that protect assets (ability to fail safely)
- Influence the ability to maintain positive control of the moderation of hazards at CRITICAL STEPS (e.g., **Hu** Tools and at-risk practices)

DO YOU NEED A PARACHUTE TO SKYDIVE?

Do you need a parachute to skydive? Only if you intend to skydive twice. With all humor aside, people do perish from skydiving. Yet it's safer than driving your car to the grocery store. The sport of skydiving continues to improve its safety record. In 2014, the United States Parachute Association (USPA) recorded 24 fatal skydiving accidents in the U.S. out of roughly 3.2 million jumps. That's 0.0075 fatalities per 1,000 jumps—among the lowest rate in the sport's history. Tandem skydiving (two people tethered, jumping together) has an even better safety record, with 0.003 student fatalities per 1,000 tandem jumps over the past decade.[3] According to the National Safety Council, a lightning strike or a bee sting is much more likely to kill a person.

With 14 fatalities, 1961—the first-year records were kept—stands as the year with the fewest skydiving fatalities. However, USPA membership was considerably smaller then, with just 3,353 members, and the total number of jumps was far fewer than today's 3.2 million-plus jumps. To put this in perspective, the 1960s averaged 3.65 fatalities per thousand USPA members. In contrast, 2014 had 0.65 fatalities per thousand USPA members. And estimating about 3.2 million jumps last year, that's one fatality per 133,333 skydives. The USPA attributes improvements in skydiving safety to better equipment reliability, automation, and collaborative adherence to basic safety requirements, as well as the centralized oversight of the sport by the USPA.[4] Table 4.1 summarizes some of the practices that create not only safety but also danger.

TABLE 4.1
The Sport of Skydiving Is Inherently Dangerous.

RIAs Performed *Before* Jumping Out of the Aircraft[5]

RIAs that Create Danger	RIAs that Create Safety
• A rigger making an error while packing the parachute	• Packing main and reserve parachutes properly
• Selecting a parachute more complicated than the skydiver can handle	• Double-checking parachutes are rigged properly before closing the backpack
• Taking off in an aircraft and climbing to 13,000 feet (necessary condition for skydiving)	• Checking the integrity of gear and lines
	• Acquiring knowledge and skill to conduct the jump properly with the selected parachute
• Securing the harness improperly	• Donning and securing the parachute harness properly
• Opening the aircraft door (necessary)	
• Stepping toward the edge of an open door	• Jump master's final check that you are ready to jump

Note: This table lists the relevant RIAs for skydiving's one CRITICAL STEP: stepping out an aircraft's open door.
Note: Notice that RIAs have a dual nature—they can create safety and danger. Certain conditions must be satisfied before a person leaps out of the aircraft to survive the descent.

RIAs always *precede* the respective CRITICAL STEP. Every CRITICAL STEP is set up by one or more earlier actions that establish the conditions that create value (desired accomplishment) or harm (undesired outcome). Every CRITICAL STEP should normally have at least two RIAs—one that creates the pathway for work and a second that protects the asset during work. Additionally, there is usually a lapse of time between an RIA and its related CRITICAL STEP (natural slack). For example, to discharge a firearm, a bullet must be inserted into the chamber (RIA) before pulling the trigger (CRITICAL STEP). The fact that RIAs precede CRITICAL STEPS suggests there is time to check that proper conditions exist before using the firearm, such as making sure the safety mechanism is on or off and pointing the firearm in a safe or intended direction.

RIAs are always *reversible.* That means an RIA can be done, undone, redone, undone, redone, and so on, with no change in the state of the asset—as long as the CRITICAL STEP is not performed. Referring to the list of RIAs for skydiving in Table 4.1, notice that every action can be reversed. A parachute rigger can rig the main parachute and then decide to unpack the bag and do it over. The skydiver can don and secure the parachute, but then decide not to jump and remove it—as long as the person does not jump out the door! Figure 4.1 illustrates the linkage between preceding RIAs and their related CRITICAL STEP for the task of skydiving. Another aspect of RIAs worthy of note is that they can decrease or increase the margin of error. While in the aircraft, moving toward or away from the open door decreases or increases the skydiver's margin of error (stumbling). When driving on the highway, safety-conscious drivers move over for those stopped along the road (pedestrians inadvertently stepping into travel lane). Such moves influence the margin of error for the performer and the asset, in some cases.

Preceding RIAs

- Fly aircraft at 13,000 feet
- Fold main and reserve parachutes properly
- Rig parachute backpack pins and cords properly
- Don and secure the backpack properly
- Open the aircraft door
- Step toward the edge of open door

CRITICAL STEP

Leap across the threshold of an open aircraft door

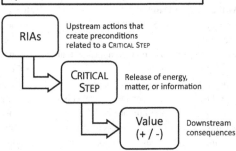

FIGURE 4.1 The linkage between CRITICAL STEP and related RIAs is illustrated for the task of skydiving. RIAs always *precede* their respective CRITICAL STEP and are always *reversible*—**if** the CRITICAL STEP has not been performed.

Source: Muschara, T. (2018). *Risk-Based Thinking* (p. 111).

In the healthcare industry, to reduce medication errors and harm to patients, healthcare workers use an **Hu** Tool: the "Five Rights of Safe Medication Practices." The five rights are the 1) right patient, 2) right drug, 3) right dose, 4) right route, and 5) right time. This technique helps healthcare practitioners recall and verify the proper preconditions *before* administering a medication to a patient, always a CRITICAL STEP. However, the five rights **Hu** Tool is not a behavioral solution for achieving medication safety. Several factors may interfere with a nurse's ability to complete these functions—recall the Nurse's experience in the Introduction. The healthcare organization must responsibly align its systems so that each right (RIA) can be accomplished.[6] Remember these *rights* the next time you're in the doctor's office.

RIA—A DOUBLE-EDGED SWORD

As noted in Table 4.1, an RIA is a double-edged sword when it comes to work execution. RIAs have a dual/binary nature in that they either establish **safety** or increase **danger** of the work activity. When performed properly, RIAs establish the preconditions to perform an associated CRITICAL STEP without harmful consequences. Performing the CRITICAL STEP is anticlimactic. However, if done improperly (including omissions), RIAs can create the preconditions for a serious event or tragedy. In other words, they can set you up for future failure, such as a landmine.

Recalling the Tourist's near-death experience with hang gliding, we illustrate how RIAs create conditions for either **safety** or **danger**. Figure 4.2 illustrates an RIA's dual character with safety and danger to help you further understand the relationship between RIAs and the related CRITICAL STEP. To successfully hang glide in tandem with an instructor requires that everything go right not only during the flight (ongoing CRITICAL STEP while in the air) but *before* the flight as well. The following is a simple, hang-gliding safety checklist:

- Helmets strapped on
- Passenger's and pilot's body harnesses buckled (secured to the body)
- Harness straps connected to the air frame, verifying all carabiners locked
- Lumbar strap attached
- Pilot and passenger hooked together (to help steer and control glider in the air)
- Hands grasping horizontal bar (at takeoff)

Note: Notice that the safety checklist is actually a check of the Risk-Important Conditions, the preconditions for safety established by RIAs. While CRITICAL STEPS add or extract value, RIAs create safety or danger related to subsequent CRITICAL STEPS.

Caution: These preconditions for safety are subject to the dynamics of work and the workplace and are worthy of continuous monitoring and conversation after RIAs, before and during CRITICAL STEPS.

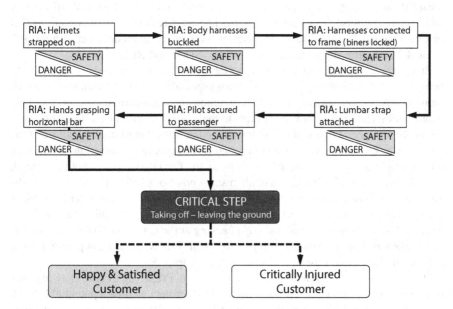

FIGURE 4.2 The dual nature of Risk-Important Actions. They either create safety or increase danger. When undetected, improperly performed RIAs create landmines, creating future CRITICAL STEPS potentially unknown to the performer.

RIAS VERSUS TOUCHPOINTS AND CRITICAL STEPS

In addition to differences between RIAs and CRITICAL STEPS, there are other general differences between RIAs and touchpoints. As stated earlier, a touchpoint is a human action that *changes* the state of an *asset* or the control of a *hazard*. RIAs are NOT touchpoints. By definition, RIAs are human actions, but they do not interact with either an asset or hazard directly to alter their state. RIAs set the stage for subsequent human actions that interact with the asset or the hazard—touchpoints. Both touchpoints and RIAs involve work—transfers of energy, movements of matter, or transmissions of information. However, the work performed by RIAs affects objects *other than* assets and hazards. RIAs do not affect the status of an asset or the state of a hazard. For example, loading a pistol with a cartridge is an RIA, but it has nothing to do with the release of energy or the state of the target/asset or state of the hazard, the bullet. Inserting a cartridge into the pistol's chamber involves work, the movement of matter (cartridge), and the transfer of energy to put it there. (The cartridge can be removed also, that is, reversible.) But the firing of the pistol, propelling the bullet out of the barrel, is controlled by the all-important touchpoint—the trigger.

Pulling the trigger is a CRITICAL STEP. All CRITICAL STEPS are touchpoints, but not all touchpoints are CRITICAL STEPS. This distinction is an important decision point in CSM (Chapter 7). Pulling the trigger is a touchpoint that changes the state of the hazard, the bullet, from stationary to traveling at 1,200 feet per second. However, most touchpoints do not satisfy the criteria of a CRITICAL STEP. Recalling the *Hu Risk Concept* (Figure 2.1 in Chapter 2), touchpoints are illustrated by the overlaps between the human and the asset and between the human and the hazard. RIAs create those overlaps. It is important to recognize these fundamental differences and reserve the term "touchpoints" for the human-directed work in operations that involve changes in the state of an asset or the control or moderation of a hazard. Table 4.2 compares the attributes of CRITICAL STEPS and RIAs.

It is crucial to validate conditions established by previously performed RIAs before a CRITICAL STEP is performed. This is especially true when teamwork involves numerous members performing different actions at various times and places. This coordination challenge can lead workers to make assumptions or use informal communication methods that can create misunderstandings and ultimately failure at the CRITICAL STEP. This is where formality, rigor, and ongoing conversations about what must go right are proven valuable. As long as the CRITICAL STEP is not performed, time is available to think and to establish safe conditions before the work proceeds. To increase reliability, we suggest using a CRITICAL STEP *hold point* and *independent verification* to verify the proper preconditions exist before performing the CRITICAL STEP and using *peer-checking* and *self-checking* at the CRITICAL STEP to ensure the right action is done to the right thing at the right time with the right preconditions. (These **Hu** Tools, and others, are described in Chapter 5.)

This sad but true tragedy did not need to happen. Preparing to fail safely requires us to anticipate a loss of control of a hazard or to protect against a potential hazard, including recoverability from the onset of harm. For Bill, traveling several miles from home was always a CRITICAL STEP. The hazards for Bill were respiratory

TABLE 4.2
Comparison of the Attributes of Risk-Important Actions and CRITICAL STEPS.

Risk-Important Actions	CRITICAL STEPS
Significant human interaction that: • Creates pathways for work, which means the RIA precedes the CRITICAL STEP • Creates "line of fire" situations • Creates necessary Risk-Important Conditions for a subsequent CRITICAL STEP • Requires high degree of monitoring and control to verify proper conditions are established before performing the CRITICAL STEP • Influences the control of hazard and/or barriers and safeguards for an asset during work	**Significant human interaction that:** • Adds value if performed properly • Necessitates positive control of hazard • Introduces potential unwanted change in state of an asset if performed improperly • Creates a *continuous CRITICAL STEP** while in a "line of fire"
Human Action: An RIA is a human action but not a touchpoint; RIAs change the condition of other objects (work), but not the asset or hazard	**Touchpoint**: Human action that changes the state of an asset or the moderation of a hazard through some work; CRITICAL STEPS are touchpoints
No Response. No system/component reaction most of the time; in some situations, there may be a system response, but would proceed slowly enough to allow people to respond to avoid an unwanted outcome (e.g., deploying the emergency parachute during descent after jumping out of an aircraft)	**Prompt Response.** Immediate in most cases; there may be a delay, but past the point of no return (falling off a cliff)
No Harm to an Asset: Work is involved to establish conditions (pathways) for the work to be done at the related CRITICAL STEP; only hazards cause harm	**Harm**: Meets the criteria of "intolerable" as defined by management; due to an uncontrolled transfer of energy, movement of matter, or transmission of information
Reversible: Option exists to start over if CRITICAL STEP has not been performed	**Irreversible:** No option to start over; no undo; past the point of no return
Margin of Error: Two or more mistakes, malfunctions, or acts away from harm	**Margin of Error**: One action away from harm
Latent Error: Human actions, verbal expressions, or decisions that establish latent conditions; not self-revealing (hidden); can create landmines	**Active Error**: Human error that triggers immediate harm to an asset, self-revealing, if there is a loss of control

Note: Every CRITICAL STEP has related RIAs that either create the conditions for the CRITICAL STEP or influence the state of related defenses (controls, barriers, and safeguards) for relevant assets and hazards.

---

* *Continuous CRITICAL STEPS* occur when a human activity persists for more than a few moments, requiring vigilance for minutes or hours, where a loss of control would result in intolerable harm, such as driving a vehicle at high speeds on a crowded thoroughfare. A continuous CRITICAL STEP exists while in the line-of-fire—no control of objects exists in the line-of-fire.

A DEADLY OVERSIGHT

It had been a perfect day of skiing in the mountains for Joe and Bill, a pair of backcountry skiing enthusiasts. They had spent the day backcountry skiing in untouched powder. Tired and hungry, they stopped in the small town for a cold beer and pizza at a local eatery, as they had done many times before. The drive home through the mountains was uneventful. However, as they approached the outskirts of town, suddenly Bill had difficulty breathing.

Bill pulled the vehicle over, as he could no longer drive safely. He recognized that he was having an allergic reaction and made a desperate call to his wife to bring his emergency medication. Bill normally carried a rescue inhaler but had forgotten it this day. Bill's wife answered the phone. She immediately understood the need for a rapid response. With the rescue inhaler in hand, Bill's wife raced out the door, with his daughter still on the phone, trying to reach her husband in time. As Bill's airway closed, his last words to his wife and daughter were, "Goodbye, I love you." They arrived too late. Bill died in a lonely parking lot as his friend Joe looked on, unable to prevent the inevitable.

irritants (dust, pollen, food preservatives, etc.) that existed broadly in life, landmines in his world. Without his inhaler, he was unprepared for an acute onset of respiratory distress, which could arise without warning. On most days, Bill carried his rescue inhaler on his person, a daily RIA before leaving his home. But on this fateful day he did not, which precluded his ability to recover from an allergic reaction. Joe, aware of Bill's health condition, did not remind Bill as they headed out. Unfortunately, Bill paid the ultimate price for his omission. The importance of getting an RIA right cannot be overstated; it may be the difference between life and death.

EXAMPLES OF RIAS IN EVERYDAY LIFE

To help the reader better understand the relationship between a CRITICAL STEP and its related RIA(s), several examples from everyday life are provided in Table 4.3. For the examples listed, the related RIA precedes the CRITICAL STEP. Also notice that in every case an asset's condition does not change—there is no work done on the asset, and the action is reversible *as long as* the related CRITICAL STEP is NOT performed.

RIAs are indeed important to safety, but people sometimes confuse these actions with CRITICAL STEPS because they rightfully perceive RIAs to be important to safety or quality. For example, many people consider securing their seatbelt before driving their automobile to be a CRITICAL STEP. But, really, is it? As long as the car is not moving, nothing happens if the seatbelt is buckled or unbuckled—no harm occurs. CRITICAL STEPS always involve transfers of energy, movements of matter, or transmissions of information that could trigger immediate harm to assets if performed out of control. Although important, such actions trigger no harm until later during actual CRITICAL STEPS.

TABLE 4.3

Examples of Risk-Important Actions and Their Subsequent CRITICAL STEPS.

Risk-Important Action		CRITICAL STEP
Adding new motor oil to an engine (after draining the old oil and tightening the oil pan drain plug) . . .	**before**	turning the ignition key to start the engine.
Performing a no-voltage check . . .	**before**	uncoupling the electrical wires to a wall socket.
Typing an email message with personal financial information . . .	**before**	clicking Send.
Slowing the speed of your vehicle below posted speed limit . . .	**before**	entering a highway construction zone when workers are present.
Marking the correct spot on a masonry wall . . .	**before**	pulling trigger to start drill with bit against the wall.
Donning a hardhat, eye and ear protection, and gloves . . .	**before**	walking into an active construction zone.
Turning the wheels into the curb after parking a vehicle on a steep incline . . .	**before**	stepping out of the car.
Grasping a handrail . . .	**before**	walking down a stairway.
Checking container's cap is securely closed with a mallet . . .	**before**	picking up a partially filled paint bucket by its bail.
Securely closing a water supply valve . . .	**before**	disconnecting the fill hose to a toilet tank.
Looking both ways . . .	**before**	stepping onto a street surface from the curb.
Opening a hamburger purchased at a drive-through fast-food restaurant to inspect its contents . . .	**before**	taking a bite of it (for those allergic to cheese).
Installing firewall software on your home computer . . .	**before**	connecting to the Internet.

Note: Notice the common word, *before*. RIAs always precede their respective CRITICAL STEP.

KEY TAKEAWAYS

1. Every CRITICAL STEP possesses one or more RIAs that establish the conditions for adding value (success, when done properly) or extracting value (harm, if done improperly). They create preconditions for safety or establish the conditions for danger.
2. The key distinctions between a CRITICAL STEP and its related RIAs are:
   a. An RIA for an associated CRITICAL STEP always precedes the CRITICAL STEP.
   b. The results of an RIA are reversible, while they are irreversible for a CRITICAL STEP.
   c. RIAs alone do not influence the state of the asset or the release of a hazard associated with the asset.

3. A check of the preconditions established by RIAs offers an opportunity to *prove* it's safe to perform a CRITICAL STEP. As long as the CRITICAL STEP is not performed, time (natural slack) is available to think and take recovery actions before the work proceeds.
4. If omitted or performed improperly, RIAs can create landmines—hidden CRITICAL STEPS later in the work activity. RIAs performed earlier in a procedure can establish unsafe conditions that may not be detected at the time a CRITICAL STEP is performed.
5. RIAs can reduce/increase the margin of error. Error is unacceptable at CRITICAL STEPS, while errors at RIAs have little or no immediate consequence, though extremely important at the CRITICAL STEP.
6. RIAs include all human actions that:
   a. Create pathways, whether via the asset or the hazard, or
   b. Increase or decrease the margin of error, or
   c. Establish control of hazards, or
   d. Impact the presence and/or effectiveness of defenses (barriers and safeguards) for assets

CHECKS FOR UNDERSTANDING

1. What are the five conditions that must be ascertained by a healthcare worker before administering a drug to a patient?
2. Which of the following human actions would you consider an RIA for starting an automobile engine just after an oil change?
   a. Reinstall the oil sump drain plug after draining the old oil.
   b. Add proper amount of new engine oil to the oil sump.
   c. Check the oil level using the dipstick.
   d. All the above.
3. What RIA(s) must be done before clicking SEND on an e-mail message to an important customer to review a high-dollar contract?
4. While preparing a meal for a 2-year-old toddler, a parent answers a telephone leaving the handle of a saucepan with hot sauce extended over the edge of the stovetop. What's the problem?
5. Before leaving on a road trip for the holidays, what actions should you take to increase the chances of an uneventful drive to your destination?

(See Appendix 3 for answers.)

THINGS YOU CAN DO TOMORROW

1. Build in hold points in operating procedures at known CRITICAL STEPS to verify the necessary preconditions are established before proceeding (outcomes of RIAs).
2. For specific high-risk operations, pick out one or more CRITICAL STEPS to lead a discussion with a work group on the RIAs that influence the safe performance of the respective CRITICAL STEP. Be certain to contrast the key distinctions between CRITICAL STEPS and RIAs.

3. Set expectations for leaders to observe and discuss RIAs and CRITICAL STEPS with frontline personnel to confirm proper controls are in use to avoid losing control and/or minimizing harm.
4. During safety meetings, training, weekly staff meetings, etc., promote a conversation on how RIAs prove safety exists.
5. If questions or doubts exist regarding CRITICAL STEPS or their related RIAs, include support group contributions to establishing and maintaining these conditions.
6. As a general practice, use the phrase "risk-important action" early and often during the implementation of CRITICAL STEPS. Avoid using the abbreviation "RIA" until people understand the concept.

REFERENCES

1 Churchill, W. (1964). *The River War: An Historical Account of the Reconquest of the Soudan.* New York: Award Books.
2 Gorgan, E. (2018, November 27). *First-Time Hang Glider Hangs on For Dear Life as Pilot Forgets to Strap Him In.* Retrieved from: www.autoevolution.com/news/first-time-hang-glider-hangs-on-for-dear-life-as-pilot-forgets-to-strap-him-in-130479.html.
3 Information acquired from the United States Parachute Association Website. Retrieved from: https://uspa.org/Discover/FAQs/Safety.
4 Ibid.
5 Information was acquired from the United States Parachute Association Website (see www.uspa.org). Table 4.1 summarizes a few of the key actions skydivers perform before exiting the aircraft.
6 Smetzer, J. (2007, January 25). 'The Five Rights: A Destination Without a Map.' *ISMP Medication Safety Alert,* 12(2). Institute for Safe Medication Practices. Retrieved from: www.ismp.org/resources/five-rights-destination-without-map.

5 Performing a CRITICAL STEP

Humans must remain in control of their machinery at all times. Any time the machinery operates without the knowledge, understanding, and assent of its human controllers, the machine is out of control.[1]

—**H.C. (Hop) Howlett Author: Industrial Operator's Handbook**

Fire in the hole!

—**Unknown origin**

BURNING ASSUMPTION[2]

Two experienced maintenance mechanics attempt to replace the packing on a manual steam valve still under pressure.* One mechanic received second-degree burns on his legs, hands, and face.

Two mechanics were assigned to adjust the packing on two steam valves and to replace the packing on a third. A portion of the 150-psig steam system was tagged out by Operations to support this effort so that no steam was flowing through the piping. However, the isolated section of piping would remain pressurized until it cooled; piping was not vented when the danger tags were hung.

At the job site, the mechanics noticed the valve that needed new packing was still blowing hot water and steam through the spent packing. So, they went ahead with the other two valves and tightened the packing as needed. When they returned to the first valve, they found the valve stem dry and the valve no longer blowing steam.

Thinking the pressure had been relieved because there was no leakage past the valve's stem, the mechanics proceeded to remove the valve packing by removing nuts on the packing follower. Suddenly, the packing blew out with a burst of hot water and steam. One mechanic on the platform side of the valve was able to move out of the way of the steam plume. The second mechanic, sitting on a support girder on the other side of the valve 15 feet above the floor, had nowhere to escape. He received second-degree burns on his legs, hands, and face but fortunately did not fall to the concrete floor below.

---

* Many manually operated valves in steam piping systems require several inches of flexible cotton yarn impregnated with graphite to inhibit the leakage of steam along the stem of the valve at operating temperatures and pressures. A metal follower (or gland) attached to the body of the valve with two bolts secures the packing around the stem so that the process fluid would not blow the packing out of the valve body. Periodically the follower must be tightened to accommodate wear of the packing.

DOI: 10.1201/9781003220213-5

The order of closing isolation valves listed on the tagout had trapped steam in the pipe with none of the system vent or drain valves tagged open to relieve the pressure. Neither operators nor the mechanics checked the system to verify it had been depressurized before work began. Scaffolding had not been requested because it would have added several hours to the job.

You could say that, as experienced mechanics, they should have known better—they did not exercise *positive control* of the residual steam hazard trapped in the still-hot steam system piping. But we all make mistakes, and this one led to one mechanic suffering serious burns. Both mechanics were in the pathway of the hot water and steam; one escaped, while the other could not. The CRITICAL STEP was removing the metal packing follower nuts from the valve that allowed steam to escape, ejecting both the follower and packing from the valve body. Among several RIAs mentioned earlier, venting the pipe section before removing the packing would have purged the steam hazard. Later, in this chapter, we introduce the concept of a landmine. Just like landmines in a battlefield, workplace landmines are conditions poised to trigger harm unknown to those involved. But first let's explore what is meant by positive control and how to exercise it.

POSITIVE CONTROL

Positive control from an operator's point of view is simply stated as, "*What is intended to happen is what happens, and that is all that happens.*" Negative control, on the other hand, is just the opposite: "What is intended to happen doesn't happen, and something else happens instead." That's a concise way of describing loss of control. The aim of positive control is to accomplish an expected result while avoiding a loss of control of the built-in hazard that's doing the work. Exercising positive control is about deliberately achieving a goal—in this case, a work goal without suffering harm.

The means chosen to exercise positive control must be proven reliable. Nominal human reliability (99.9 percent) is acceptable for non-critical work. But if a person is attempting to perform a CRITICAL STEP, then human performance must be augmented with methods to exercise positive control. Methods vary but involve the application of various types of defenses. Defenses function mostly to prevent and protect.[3] *Controls* guide behavior choices and actions (preventive). *Barriers* impede the movement of energy, matter, and information (preventive and protective), and *safeguards* mitigate or recover from the onset of harm (protective). Controls, in particular, rely on a person's knowledge, skills, proficiency, character, values, and the person's choice to use them, such as following procedures; adhering to signs, rules, approvals, and regulations; responding to coaching and correction; and applying various non-technical skills, such as **Hu** Tools, described later in this chapter.

Before performing a CRITICAL STEP, the performer considers 1) the asset and its critical parameters, 2) the intended outcome (expected change in the asset's critical parameters), 3) the built-in hazard used to transform the asset and related controls, 4) the preconditions for safety to protect the asset (barriers and safeguards), and 5) the action to achieve the outcome (means of positive control). Once the performer is confident, only then is the action performed.

Let's reconsider the handling of a firearm. Responsible firearm users apply the four standard rules of gun safety, which provide a valuable analogy to how workers should approach any high-risk work containing a CRITICAL STEP[4]:

1. Always handle a firearm as if it is loaded. **Always.**
2. Always point the firearm in a safe direction. **Always.**
3. Always know your target and what is beyond it. **Always.**
4. Always keep your finger off the trigger until you are ready to shoot. **Always.**

The preceding series of rules offers insight into the moment before a CRITICAL STEP is performed—a pause for safety. Placing the index finger inside the trigger guard is an RIA, which places the finger one act away from discharging the firearm. Pulling the trigger is the CRITICAL STEP. There is a moment between placing the finger inside the trigger guard and pulling the trigger to think about the act and its consequences. Properly executing RIAs determines whether someone or something is seriously injured or damaged. This stop-and-review moment involves an **Hu** Tool commonly known as "self-checking." (See **Hu** Tools later in the chapter.)

Safety is better considered a process—it is something the organization (or individual) *does* rather than something the organization possesses, such as a VPP certification,* a training program, a quality program, procedures, or a safety analysis report.[5] Besides using barriers and safeguards, organizations enable positive control—the mindful performance of those at the sharp end. Considering this perspective, identifying and controlling CRITICAL STEPS are two things a performer can *do* to ensure safety happens to protect assets from harm during work. *Most of the time, safety is conceived in the minds of the performer and the team before work starts.* Positive control aids in detecting unsafe conditions or otherwise informing, creating awareness of, and inhibiting unsafe acts. Collectively, positive control tends to improve the reliability of successful work while reducing active errors, which tends to decrease the frequency of events.[6] However, the central weakness with positive control is that it depends entirely on the performer to recognize danger, think prudently, and respond safely. How he/she thinks is important to positive control.

EFFECTS OF FAST AND SLOW THINKING ON POSITIVE CONTROL

In Chapter 2, we introduced fast thinking (FT) and slow thinking (ST) and their roles in the *Work Execution Process.* Now we want to dig deeper regarding their

---

* Voluntary Protection Programs (VPP) is a U.S. Occupational Safety and Health Administration (OSHA) initiative.

applications to positive control. Conceptually, the time (t) performing a task is a combination of time devoted to thinking and time devoted to doing.

$t_{task} = t_{thinking} + t_{doing}$

* *Thinking*—the work of using the mind to generate thoughts; noticing, gathering information, remembering, reasoning, evaluating, deciding, and choosing; involving mostly ST
* *Doing*—accomplishing physical work (force applied over a distance); executing; achieving the production/efficiency goal(s); involving mostly FT

If the time available for the task is less than the time required to do it safely and reliably, then human nature tends to sacrifice the remaining time to doing—to stay on schedule. Time constraints and schedule pressure tempt people to shift from ST (being thorough) to FT (being efficient).[7]

Time pressure is not a bad thing—it comes with the marketplace. The issue is how the individual responds to it, such as hurrying and cutting corners. Hurrying (FT) is a dangerous choice during CRITICAL STEPS. However, you should readily recognize this well-known operational principle: *slow down to speed up*. Doing it right the first time, though slowly, is faster than working quickly and making mistakes that cost not only time but also money and maybe lives. Later, in this chapter, we introduce **Hu** Tools, which help combat the human tendency to work fast and efficiently. The use of **Hu** Tools builds in time where it's needed most, to exercise RISK-BASED THINKING.

ST deliberately focuses the mind's attention on something. Unlike FT, ST must be turned on. In the workplace, the performer must turn on attention to learn, anticipate, monitor, and respond, all required to perform a CRITICAL STEP. Vigilance is necessary to detect unsafe conditions and surprise CRITICAL STEPS, but it cannot be maintained indefinitely. The mind tires and it becomes impractical to remain alert for prolonged periods. Therefore, when we feel it's safe, we turn off ST.

FT minimizes effort and optimizes performance. It's unconscious and learned over time. Malcolm Gladwell, in his popular book *Blink*, refers to FT as the "efficient thinker." It's fast and frugal.[8] In some ways, FT is related to skill-based performance, which is performance guided by muscle memory and habit developed over time by repetition and experience. FT responds to a current mental model that the performer believes represents what is normal and current for the situation: beliefs about how things work—cause and effect. Thinking in fast-paced yet dangerous situations can be trained to respond reliably, such as downhill ski racing. The FT brains of frontline workers can be educated and influenced to recognize threats of danger by regularly scheduled refresher training that develops their technical expertise (enhancing expert intuition) and by instilling a chronic uneasiness toward pathways for harm—impending transfers of energy, movements of matter, and transmissions of information.[9] In this respect, FT can be a valuable ally to ST![10]

At times, one or the other, FT and ST, is appropriate. But you must consider the risk at hand when choosing to transition between the two. When driving an automobile, we tend to drive as fast as we desire, depending on our perception of risk of losing control and crashing or getting a citation from a law enforcement officer. We speed up when we think it's safe and slow down when it's not. When do you shift? Surprise is an occasion when situations suddenly don't match expectations based on your understanding of how things work—your previous mental model, which has just now been challenged. Surprise activates and orients attention, turning on ST, especially when a pathway is recognized or other triggers of danger (audible, visible, feeling that something is not right) are revealed. Besides surprise, there are other occasions to shift to ST. Table 5.1 provides examples of situations in which either FT or ST would be appropriate.

Dr. Daniel Kahneman concludes that FT is the source of much of what we do wrong.[11] Human beings tend to jump to conclusions when thinking fast (habitual, emotional, instinctive responses). You miss details when engaged in FT. The performer must rely on the internal risk monitor—expert intuition—to sound the alarm for danger while working efficiently. Generally, conclusions reached unconsciously are based on faith that "What you see is all there is" (otherwise known as mindlessness). Dangerous in high-risk situations! We hope it is obvious that FT is inappropriate for CRITICAL STEPS—it's unsafe!

The next section is devoted to the use of **Hu** Tools, which exercise various forms of RISK-BASED THINKING, all requiring ST. **Hu** Tools help the performer know when to shift from FT to ST to slow down work to take time to think.

TABLE 5.1
Examples of When Either Fast Thinking or Slow Thinking Are Appropriate.

Fast Thinking	Slow Thinking
• Walking and running	• At *CRITICAL STEPS*, e.g., stepping out the door of an aircraft flying at 13,000 ft
• Survival instincts (freeze, fight, or flight responses)	• At *RIAs*, e.g., donning a parachute before jumping out of the aircraft
• Riding a bicycle	• Any significant transfers of energy, e.g., striking a match near the pilot line on a gas water heater
• Playing a musical instrument	• Substantial movements of matter, e.g., lifting a heavy object or adding gasoline to an automobile
• Writing one's signature	
• Operating a toaster	• Important transmissions of information, e.g., depressing the enter key on a computer keyboard
• Boarding a bus or people-mover	
• Typing on a computer keyboard	• Entering a line-of-fire, e.g., stepping into a street
• Opening/closing a water faucet	• Conducting exchanges of value, e.g., clicking "purchase" on a retail Internet website
• Flushing a toilet	
• Washing your hands	• Changes in risk situation, e.g., changes in traffic patterns while driving a vehicle
• Driving a vehicle	
• Water or snow skiing	• Revising software settings, e.g., revising the settings on a Smartphone
• Operating heavy equipment	
• Swerving to avoiding an obstacle in the roadway	• Planning a project or preparing for an excursion, e.g., a road trip, a boat ride, or a backcountry adventure

Note: A CRITICAL STEP is a prime example of when slow thinking is necessary. Fast thinking at a CRITICAL STEP is dangerous.

HUMAN PERFORMANCE TOOLS PROMOTE POSITIVE CONTROL

A Human Performance (**Hu**) Tool is a set of discrete mental skills and behaviors that help workers perform their activities more reliably, minimizing the chances of losing control during work. **Hu** Tools apply one or more mental practices—ST—associated with RISK-BASED THINKING (anticipate, monitor, respond, and learn). As defensive controls, they guide behavior choices. **Hu** Tools are considered *skill-of-the-craft* and are not proceduralized. **Hu** Tools, when used mindfully and rigorously, do the following:

- Enhance positive control of human actions, promoting anticipation, monitoring, predictability, and learning
- Aid decision-making regarding in-field adjustments, responding to emergent workplace conditions that threaten assets—to fail safely

Hu Tools help workers accomplish both purposes; but in either case, the individual must think mindfully and intentionally. Using **Hu** Tools is an explicit acknowledgment of the presence of risk and the need for caution. They can be applied as needed anywhere in a work activity. Using **Hu** Tools takes time—some more, some less; every tool is designed to slow down the tempo of performance to give you time to think, RISK-BASED THINKING.[12] Earlier, we described the RU-SAFE prework discussion method, a preparation phase **Hu** Tool. The **Hu** Tools described in the following apply mostly to the execution phase of work.

> **Caution**: RISK-BASED THINKING is ST, but ST is not necessarily RISK-BASED THINKING. Top performers are characteristically uneasy in that they maintain situation awareness (always learning), habitually anticipate risk, seek and provide feedback, monitor pathways and an asset's critical parameters, and take actions to proceed cautiously and methodically. These are examples of RISK-BASED THINKING that require ST. ST associated with checking scores of your favorite sports team on your mobile device is not RISK-BASED THINKING.

SELF-CHECKING

Self-checking is one of the most effective **Hu** Tools an individual can use to exercise positive control at a CRITICAL STEP.[13] Once mastered, the technique takes only seconds to use. Self-checking applies all four aspects of RISK-BASED THINKING (anticipate, monitor, respond, and learn). Self-checking involves stopping the flow of work momentarily to help the performer concentrate, to anticipate what is about to happen and what critical parameters to monitor—deliberate ST.

Before transfers of energy, movements of matter, or transmissions of information, the performer pauses to focus attention, followed by a moment to reflect on what is to be accomplished and what to avoid. The performer anticipates whether the proposed action is the correct action on the right component at the right time, what should happen after taking the action, and what to do if an unexpected result occurs (contingency to fail safely). Self-checking's effectiveness depends greatly on the performer's grasp and understanding of the transformation process—his or her technical knowledge and expertise. If uncertain, he or she resolves any questions or concerns before proceeding (see Stop When Unsure).

When confident of the plan, the action is performed. While monitoring the asset's critical parameters, the performer confirms the expected outcome and the change in state. If the expected outcome is not evident or is different than anticipated, the person stops work, placing everything in a safe condition, if practicable. Then the performer either implements a contingency or gets help to verify the true conditions before proceeding with the work. The memory aid **STAR** (stop, think, act, review) helps the user recall the thoughts and actions associated with the means of self-checking.

1. *Stop*—Just before performing a CRITICAL STEP, pause to:
   a. Eliminate distractions.
   b. Focus attention on the asset and its hazard—at the pathway.
2. *Think*—Understand what will happen, especially to an asset, when the action is performed.
   a. Verify the action is appropriate, given the status of the asset(s); understand the pathway for the transfer of energy, movement of matter, or transmission of information.
   b. Recall the critical parameters of the asset to monitor, how they should change, and the expected result(s) of the action.
   c. Consider contingencies to minimize harm if an unexpected result occurs.
   d. If there is any doubt, STOP and get help. Otherwise, apply STOP-work criteria.
3. *Act*—Touch the correct control; and without losing physical contact or attention, perform the correct action on the correct component.

   Note: It's common practice in some hazardous situations to announce the impending CRITICAL STEP—*callouts**—to others nearby in clear, unmistakable terms, such as:

   • "Clear!" (before shocking a patient in cardiac arrest with a defibrillator)
   • "Fire in the hole!" (before detonating an explosive)
   • "Dive! Dive!" (before submerging a submarine to be alert for leaks or flooding)

4. *Review*—Verify that the expected result is obtained.
   a. Verify the desired change in critical parameters of the asset.
   b. Stop work, if criteria are met, and notify a supervisor or those with expertise.
   c. Perform the contingency if an unexpected result occurs.

---

* Callouts are commonly used in the commercial aviation community when performing a checklist. Usually, the callout is acknowledged by a second pilot as part of a double (peer) check.

Japan's train industry enhanced self-checking's effectiveness with a technique called *shisa kanko* ("pointing and calling").[14] The technique involves making large gestures with their hands and loudly speaking the status, thereby heightening focus and attention. A 2011 Osaka University study of train conductors revealed that when asked to perform a simple task, workers made approximately 2.5 mistakes per 100 actions. However, when using *shisa kanko*, the number of errors dropped surprisingly to near zero.[15] The pointing and calling technique has been scientifically validated to reduce error rates, improve recall, and, in some cases, increase the speed of accurate decisions.

TV-STAR is an American version of *shisa kanko*, which enhances the aforementioned self-checking technique by emphasizing two actions associated with STAR: touch and verify.

1. *Touch*—Physically touch the component or label or hover the computer's cursor over the component/device you intend to manipulate. The physical act of touching may activate expert intuition. Caution: the performer should self-check again if physical contact is lost or attention wanders, especially if there are several similar components in proximity.
2. *Verbalize*—Without losing eye contact, say aloud the component name or label; compare with the guiding document. Speaking uses a different part of the brain, improving awareness of the action at hand. If being peer-checked, speaking must be audible to the peer, providing an opportunity for that individual to intervene, if necessary.

Occasionally, verbalizing the component nomenclature is augmented with a peer-check. This practice is sometimes referred to as *correct component verification*, which involves a concurrent, second check by another qualified individual to verify* the performer is manipulating the proper control or equipment.

PEER-CHECKING

Peer-checking enhances positive control using a second person, a form of double-checking, except that the second person does the double-check. The worker (performer) and the checker (peer) agree that the action to be performed is the correct action on the correct component with the right preconditions. Peer-checking augments self-checking but does not replace it.

Peer-checking involves close monitoring of a performer's actions by a second technically knowledgeable person who is prepared to catch and avoid a loss of control by the performer.[16] This tool takes advantage of a fresh set of eyes not distracted by the performer's primary task. A peer may see things the performer does

---

* There is a convention for use of the verbs *verify* and *check*, especially when used in procedures. For consistency's sake, *verify* is reserved for authenticating the condition of equipment before or after performance, while *check* is associated with ensuring the performer's actions are indeed the proper actions.

not see. Peer-checking is also relevant for RIAs that create hidden preconditions for CRITICAL STEPS that cannot be verified later. For example, rigging (folding and stuffing) a parachute properly is especially important for its deployment after someone jumps out of an aircraft (CRITICAL STEP). But the jumper cannot verify the rigging of the parachute because it was laced up long before its use on the jump. It helps to peer-check as the chute is laced up, to confirm that the rigger folds and stuffs the parachute properly. A common peer-check protocol entails the following actions:

1. *Performer and the peer agree* on the intended action, on which component, and the intended result.
2. *Peer prepares to stop the performer* to avoid a loss of control and is in a position to see and hear.
3. *Performer executes the intended action* on the correct component, while the **peer** closely monitors the performer to ensure the proper action is taken on the correct component.
4. *Performer and peer confirm the expected results* after the action.

Frontline workers should be encouraged to ask for a peer-check when they believe the risks or conditions warrant it, especially for a CRITICAL STEP. Asking for a peer-check is not an admission of incompetence, but an acknowledgment of one's fallibility. A healthy culture that values peer-checking also welcomes a peer-initiated check rather than waiting to be asked. If a person other than the performer anticipates that a worker's action may be critical, unsafe, or otherwise at risk—especially if the asset's defenses are bypassed—he or she should challenge the performer to understand the intent and desired result before the action is taken, without the performer taking offense at the question.

HOLD POINT

It is advantageous to pause—take a timeout—when a pathway has been established. An established pathway means a hazard is poised to do work on an asset. This is a good time to confirm that 1) the work can be performed without losing control, and 2) the asset will not suffer serious harm if the performer loses control—that is, fail safely. This does not refer to a quality-related hold point as it is traditionally known. This is a hold point for CRITICAL STEPS. The primary purpose of a CRITICAL STEP hold point is to verify the proper preconditions exist to allow the safe performance of the CRITICAL STEP without harm—an anticlimactic result for a high-risk operation. In most cases, the preconditions were established by RIAs.

Hold points scheduled for known CRITICAL STEPS are usually guided with a checklist of preconditions to be verified, like a pilot's checklist prior to different phases of flight, such as takeoffs and landings. When conducting high-risk work in a team context, the effectiveness of the hold point is enhanced through face-to-face dialogue. The timeout also offers everyone the opportunity to validate their situation awareness of the operation at hand, ensuring they all have an accurate mental model of what is indeed happening.

INDEPENDENT VERIFICATION

Independent verification (IV) is a series of actions by two individuals (performer and verifier) working *independently* to confirm the condition of a component after the original act that placed it in that condition. IV is particularly suitable for the verification of the conditions established by earlier RIAs. IV confirms the condition of equipment (including energy, matter, or information) required to be in a particular condition to maintain the component's physical configuration required for later operation. Otherwise, adverse consequences could result later if the improper condition remains undetected—a landmine. IV can only be used when an immediate, adverse consequence of a performer's mistake cannot occur, because IV catches the *results* of errors after they have been made, not before or during error commission. IV does not work at a CRITICAL STEP but works best for verifying the preconditions established by RIAs.

To be effective, IV necessarily requires independence. True independence requires *separation in time and space* between the performer and the verifier. IV tends to have a higher probability of catching a defect than peer-checking, because the verifier's knowledge of the system, component, or work situation is unaffected by the performer. A peer-check is not independent. This practice ensures *freedom of thought* for the verifier. True independence cannot exist if one individual is looking over the shoulder of the other, even from a distance. The primary advantage of rigorous IV is the higher assurance of actual conditions before a later CRITICAL STEP. IV's disadvantages include additional time, exposure, staff availability, and costs necessary to use IV regularly.

PROCEDURE USE

Technical procedures direct the actions of a user for a task in a predetermined sequence and minimize reliance on a worker's memory, avoiding ad hoc choices made in the field. A written set of instructions improves the repeatability and predictability of complicated work activities. For complex processes, procedures also play a significant role in maintaining system configuration assumed by the procedures' authors.

Most technical procedures are based on the expert knowledge and years of experience of the authors; most are reviewed, verified, and validated extensively; and most will be accurate and up to date. Frontline workers should follow them carefully but mindfully. However, following procedures to the letter is not inherently safe. Even the best procedures are based on assumptions about workplace and system conditions. Occasionally, they have embedded inaccuracies in them due to the dynamic nature of the work and the workplace. Therefore, frontline workers should never consider procedures to be flawless. No procedure is a substitute for human intelligence and mindfulness—an act of ST. Proper use of technical procedures that contain CRITICAL STEPS adopt the following mindful practices:

1. When working with paper copies, verify the procedure is the latest approved revision.

2. Review the procedure prior to starting the work to confirm that prerequisite and initial conditions are met, that there is a match between the procedure and current conditions, and that limits and precautions (assets and hazards) are understood.

3. Keep the procedure in the user's presence in the workplace for ready reference. If working with multiple work locations, ensure each team location has ready reference to the procedure and are aware of steps/actions completed and those in progress.

4. Read and understand each step before performing it.

5. Use a place-keeping method to avoid missing a step, duplicating a step, or otherwise performing steps out of sequence.

6. Follow directions as written in the sequence specified, without deviating from the original intent and purpose unless approved otherwise by a technical authority.

7. Complete each step before starting the next step.

8. Do not mark a step "N/A" (not applicable) unless approved by a technical authority.

9. If the written instructions cannot be performed as written, are technically incorrect or confusing, will damage equipment if performed as prescribed, will result in exceeding one or more of an asset's critical parameters, or are otherwise unsafe, stop the task, place the process system in a safe, stable condition (stop transferring energy, moving matter, or transmitting information), and contact a supervisor (or persons with appropriate knowledge).

10. If the desired or anticipated results are not achieved after performing a step, stop the task, place the process system in a safe, stable condition, and contact a supervisor (or persons with appropriate knowledge).

11. Review the document at the completion of the task to verify that all steps were completed.

Remember, mindfully following approved technical procedures is one of the safest things an operator or technician can do.

THREE-PART COMMUNICATION

Three-part communication involves a repetitive verbal exchange of information between two or more people. Information related to equipment or product status or personnel protection is of primary importance. All communications during work, especially at CRITICAL STEPS and their RIAs, should be formal, clear, concise, and free of ambiguity. The repetition of the information enhances the reliability of the transfer and understanding of information. This practice is crucial with all forms of verbal exchanges—whether face-to-face, via telephone, or by radio. Throughout three-part communication, we strongly suggest using standard nomenclature, augmented with a phonetic alphabet for the English language. All parties must speak a common language.

Simply put, communication is a message sent and message received and understood. To ensure mutual understanding in three-part verbal communications involving high-risk operations, each person (sender and receiver) has specific responsibilities in the exchange.

1. *First*, the sender gets the attention of the receiver, using the person's name or title. Then the sender speaks the message clearly and concisely.
2. *Second*, the receiver repeats the message back in his or her own words. Paraphrasing the message by the receiver helps the sender ascertain that the receiver understands the intended message. However, equipment names, identifiers, and data should be repeated back exactly as spoken by the sender. If the message is lengthy, the receiver should write it down.
3. *Third*, the sender acknowledges that the receiver heard and understood the message.

The third step in the communication is the weak link, because the sender is tempted not to pay attention to the receiver's statement, assuming the person heard the message. Hearing does not always equate to understanding. If the receiver does not understand the message, he or she should not be shy to ask for clarification or for the sender to repeat or rephrase the message. Alternatively, the sender corrects the receiver, restating the message—the sender is responsible for ensuring the receiver understands the message.

CONSERVATIVE DECISION-MAKING

Conservatism means placing the safety and integrity of the asset ahead of the business goal. When an asset's safety is doubtful, frontline workers must defer to safety, not production. When safety and production goals conflict, the asset's safety is the default response.

> **Caution**: The less time available to make an informed decision, the more readily one should default toward safety of high-value assets regardless of production goals.

If in an emergency, a structured, forward-looking thought process must be used to guide decision-making, especially for situations that have no apparent guidance. Preferably, the decision-making process has been practiced in a training context such that workers have become proficient with it in a stressful situation. The memory aid **GRADE** helps in recall of the following five steps.

1. *Goal*—Identify the desired safety state of the asset in question, pinpointing the critical parameters that would be used to validate the desired safety state.
2. *Reality*—Recognize the current safety state of the asset under threat of harm, again using critical parameters and acknowledging the interaction with specific hazards, that is, pathways.
3. *Analyze*—Understand the difference between the goal state and the current state and develop means to restore the asset to the goal safety state.
4. *Decide and Do*—Choose and execute the course of action that will achieve the goal safety state.
5. *Evaluate*—Monitor changes in the asset's critical parameters to verify recovery to the goal safety state.

Conscious doubt and uncertainty are within the domain of ST, NOT FT. If you don't have doubts, you haven't been thinking! Doubts surrounding Critical Steps must be resolved before proceeding—STOP when unsure!

Stop When Unsure

If in doubt, there is no doubt. STOP! **Never proceed in the face of uncertainty** when performing a Critical Step! When people are confused, uncertain, doubtful, or confronted with audible or visible cues that something is not right, the chance for losing control increases dramatically.[17] If the proper preconditions for the Critical Step have not been established, then do not perform the Critical Step. If the RIAs were performed properly, conditions should allow the Critical Step to be performed with no surprises.

This tool presumes the worker has checked all process instrumentation or field sources of information before stopping work. The delay allows everyone involved, including the supervisor and those with relevant technical knowledge, to resolve an issue before resuming work.

1. *Stop and Secure*—Stop all transfers of energy, movements of matter, and transmissions of information, and place equipment and assets in a safe, stable condition. Take your hands off the equipment and physically back away from the work.
2. *Involve*—Consult co-workers and team members, preferably subject-matter experts. Get the facts!
3. *Notify*—If still unsure, speak with your supervisor, team lead, or respected technical authority.
4. *Solve*—Resolve the problem conservatively before resuming work. (See conservative decision-making.)

Stop When Unsure (**SINS**) provides an overarching reminder that in any situation—whether related to a written procedure or not—in which a worker is unclear on what to do or on the result of the action, he or she is obligated to **STOP** work until the issue can be clarified. Consequently, an individual is wise to stop and get help from others who have relevant knowledge. **Never gamble with safety!**

WHEN GOOD ALLIGATORS GO BAD[18]

Feeling uneasy about wrapping your arms around a 10-foot-long, 350-pound alligator would come naturally to most people—but not Kenny Cypress. The veteran alligator wrestler had entertained thousands of spectators with the Florida reptiles most notably by inserting his head into the open jaws of a wild alligator.

Kenny knew the risk and how important it was to make sure no foreign object touched the alligator's tongue. The slightest sensation inside the mouth of an alligator would trigger an instinctive reaction for the animal's jaws to snap shut. Kenny had performed this act hundreds of times before. But this time, Kenny made a mistake. After performing the show three times earlier that same day, on a hot, humid New Year's Day, perhaps the act had become so routine to Kenny that he forgot something. Though he wiped the sweat from the left side of his forehead (with the shoulder of his shirt), he forgot to wipe the right side (no self-check).

A helper in the gator pit did not check or remind Kenny to wipe his brow (no peer-check). Then it happened. A drop of sweat—unavoidable on a hot, muggy day—dripped onto the gator's tongue. Immediately, the gator's jaws locked down on Kenny's head in a potential death grip. Two hundred horrified tourists watched four men struggle for what seemed like an eternity to pry open the alligator's mouth. They were not prepared! Kenny survived, but not without scars and some permanent hearing loss.

As a circus act, inserting one's head into the gaping jaws of an alligator—in the line of fire, so to speak—is designed to provoke awe, wonder, and terror in a paying audience. The circus act purposely created a dangerous situation, which seems to most people to be a foolhardy thing to do, as is *taming* lions and tigers in a cage. Although the organization, especially Kenny, was aware of the hazard, it was not prepared (no contingencies). No tools were at the ready to extract Kenny from the gator's jaws, just in case. No one had the responsibility to ensure Kenny's forehead was wiped clear of sweat (both hands gripped the jaws of the gator) before the key act. There was no rational anticipation of what could go wrong before the "gotcha" moment. Entering the *line of fire** is, in any situation, always a CRITICAL STEP, especially when you do not have control of the movement of objects—or jaws.

Although **Hu** Tools aid in exercising positive control of high-risk operations, there are workplace conditions that conspire to seize positive control away from the performer, tending to provoke a loss of control.

ERROR TRAPS—THREATS TO POSITIVE CONTROL

Error traps are *local factors* with malevolent intent.[19] Procedures that make no sense, excessive/confusing documentation, hurrying, tools that don't fit or function, lack of technical knowledge, obscure visibility, and even oppressive bosses (among many other conditions) all conspire to prevent people from getting anything accomplished reliably without error. Error traps foster uncertainty or deviation potential in behavior, increasing the likelihood of losing control. *Variation in behavior leads to variation in results.* Error traps that provoke dangerous variation and uncertainty in workplace behavior at CRITICAL STEPS are a strategic risk.

---

*   The line of fire is the physical path that energy or an object would travel if stored energy were released uncontrolled; an asset is exposed to the potential for harm while in the path.

Most error traps involve a mismatch between the task and the limits of human capability. For example, as the speed of performance increases, the accuracy of results tends to decrease. Error traps have been otherwise referred to as "error precursors" and "performance-shaping factors."[20] Examples of error traps include the following:

TABLE 5.2
Common Workplace Error Traps that Tend to Provoke a Loss of Control.

Common Workplace Error Traps*

• Hurrying	• Unfamiliarity with or first	• Inexperience or lack of
• Competing goals	time performing a task	proficiency with a task
• High workload	• Nearness to achieving a goal	• Lack of knowledge
• Vague procedure	(summit fever)	• Lack of skill with a task
• Distraction	• Interruption	• Unclear expectation
• Multiple, concurrent tasks	• Debilitating fear	• Out-of-service
• Change from routine	• Stress and fatigue	instrumentation
• Schedule pressure	• Habit	• Confusing terminology
• Overconfidence	• Improper tool	• No equipment label

* This is not a complete list.
Note: These adverse local factors threaten the exercise of positive control at CRITICAL STEPS.

Although error traps tend to increase the likelihood of error, this doesn't mean that a loss of control will occur. An increase in error traps tends to increase the chance for error. Decreasing the number of error traps tends to increase the likelihood of retaining positive control.

If error traps cannot be designed out of the work or otherwise eliminated, it's a good practice to inform frontline workers of their presence, especially for CRITICAL STEPS. In commercial aviation, experience with crew resource management has shown that understanding the types of local factors that increase the likelihood of error enables a flight crew to remain wary of the potential for error. Such awareness reinforces the need for vigilance, wariness, and disciplined monitoring and cross-check strategies between pilots during critical phases of flight. Researchers concluded that a firm grasp of prevalent error traps on the flight deck and the related conversations assist pilots in avoiding critical errors.[21] Knowledge of error traps in general can trigger knowledge of specific error traps on the job, local factors that could trigger a loss of control at an inopportune time at a CRITICAL STEP. The coincidence of error traps with CRITICAL STEPS is a potentially deadly situation.

LANDMINES—HIDDEN CRITICAL STEPS

No system is perfect. Nothing is always as it seems. Operations, the workplace, and its organizational system are fraught with complexity, ambiguity, change (sometimes rapid), uncertainty, limited resources, and goal conflicts—all dynamic, some temporary. Few organizations have up-to-date procedures. Equipment and tools wear out.

Landmines are accidents waiting to happen. They are hidden pathways—workplace hazards *poised* to trigger harm unbeknownst to the performer, disguised threats to safety. Landmines involve the unexpected and unintended combinations of local conditions and human performance that can result in harm.[22] Adverse conditions ("set-ups") in the workplace *combine* with normal routine worker actions to trigger uncontrolled transfers of energy, matter, or information with one action. For example, a mispositioned manual valve left open during a previous work activity creates the potential for a loss of containment during later operations when workers presume that same valve is closed as assumed by the guiding procedure.

Pulling the trigger on an "empty" firearm will not hurt anyone or damage anything. But if a cartridge is unintentionally left in the chamber, pulling the trigger on a firearm thought to be unloaded can kill someone or destroy something. An unknown source of harm (cartridge in chamber) or a missing defense (malfunctioning safety lever) is a landmine in any operation. From a bucket of paint perched on an unstable stepladder to the handle of a saucepan extending over the edge of a stove top accessible to a young child, we all recognize certain hazardous situations. A proactive organization that possesses a collective uneasiness toward weaknesses in the system will be diligent to search regularly for them as well as train frontline personnel to recognize them (for example, the Foreign Object Debris (FOD) walks* done on the flight decks of U.S. aircraft carriers). A preemptive search will turn up landmines, such as:

• Pressurized systems that should be depressurized	• Hidden equipment or system configurations
• Out-of-service or faulty process instrumentation	• Inoperable, disabled, or bypassed emergency systems
• Long-term equipment deficiencies	• Inaccessible controls
• Workarounds	• Compressed springs
• Charged capacitors	• Missing or weak machine guards and railings
• Mispositioned components	• Software errors
• Worn-out or missing tools	

And there are others! None of the above are out of the ordinary. Yet, if unexpected, they are dangerous. It is important for frontline personnel to acknowledge that such conditions aren't supposed to be there but are. Adherence to procedures and expectations is important, but it is dangerous if work is performed mindlessly. A persistent state of unease and ongoing conversations among work group members are important for discovering landmines in the workplace. When discovered, correct the condition, if practicable. If not corrected promptly, known landmines become workarounds. Fix it now!

Caution: During the execution of work, the initial conditions presumed to exist may or may not be present. It is important, therefore, for frontline workers to be wary—mindful—of the pathways that may not have been identified during planning and preparation. The combination of technical expertise, chronic unease, conversations, and expert intuition of frontline workers provides the best source of safety for the unexpected.

---

* Every morning aboard operating aircraft carriers, a few hundred people inspect the deck for foreign objects, dirt, and debris, which could seriously damage a jet engine. Officers and sailors walk shoulder to shoulder picking up anything that isn't supposed to be on the flight deck. FOD walks occur more frequently for high-tempo operations.

MODIFICATION INSTALLS A LANDMINE[23]

During a test of the backup control station at a pressurized water reactor electric generating station, three power-operated relief valves on the steam side of three steam generators (heat exchanger between reactor cooling system and main steam system) surprisingly opened. This resulted in a rapid decrease in primary system pressure and temperature and a resulting contraction of coolant inventory in the nuclear reactor.*

The plant's design bases (engineering safety case) required an alternate means of plant control if the main control room had to be evacuated. Control from a backup control panel was being tested. At the time of the test, the operator at the backup control panel was unaware that a recent modification had changed the function of the valve controls on the panel (a 10-turn potentiometer for each valve). The previous control function varied the pressure setpoint at which the valve would open to modulate steam generator pressure at the setpoint. The new control function varied the valve position as a "percent open demand."

The labeling for the controls had not been revised to reflect this change. Also, the procedure the operator used had not been revised to include directions for properly presetting the controllers for the power-operated relief valves to zero turns before transferring control to the remote panel.

When control was switched from the main control room to the backup control panel, the power-operated relief valves immediately moved to 75 percent open—all three potentiometers were at 7.5 turns. The operator saw the sudden decrease in steam generator pressure. To ensure the power-operated relief valve closed, he manually adjusted the potentiometer upward to what he thought was the pressure setpoint; the potentiometer was turned to a value higher than 7.5 turns. Instead of closing the valves, this action caused the relief valves to open even more, exacerbating the overcooling effect on reactor coolant inventory already occurring.

The shift supervisor overseeing the test stopped it when level and temperature alarms annunciated in the main control room. Control was restored to the main control room, where operators closed the relief valves and returned pressures and coolant levels to normal. No nuclear fuel was uncovered, and no damage occurred.

The landmine was the repurposed 10-turn potentiometer that was set to 7.5 turns. The operator thought the number of turns set the demand pressure at which the valve would open to limit pressure rise in the steam generator. Instead, the engineers modified the number of turns to control the percent open. He was set up! Left unchecked,

---

* There was no leakage of coolant from the reactor. The significant overcooling of the water sparked by the opened relief valves in the steam side of the steam generator increased coolant density on the reactor side, triggering a lower volume of coolant due to shrinkage, not a leak. Reducing the volume of coolant in the reactor cooling system, left unchecked, could possibly uncover the reactor fuel, inhibit heat transfer, and overheat the nuclear fuel.

the contraction of the reactor coolant could have uncovered the reactor core, allowing it to overheat and damage the nuclear fuel. The CRITICAL STEP was the transfer of plant control from the main control room to the backup control panel. The related RIA would have been to verify the potentiometers were set to zero turns before making the transfer. Fortunately, the control room was paying attention, returned control back to the control room, and shut the power-operated relief valves. You could say the test failed safely. By the way, do you recognize the organizational role in the incident?

CRITICAL STEPS ARE ALWAYS CRITICAL!

CRITICAL STEPS are always critical—whether the threat is present or not! Pulling the trigger on a firearm—loaded or not—is always a CRITICAL STEP. After ejecting the magazine of cartridges, you may have forgotten to eject the round still in the chamber. Without a clear appreciation that critical is always critical—even if nothing has ever gone wrong before—frontline workers will tend to take safety for granted. They will progressively ease pressure on themselves to stay alert to landmines or the potential harm of a CRITICAL STEP gone wrong. Routine success can erode one's wariness or sense of unease. Therefore, know the condition of equipment before acting—visually and by feeling, check there is no round left in the chamber!

KEY TAKEAWAYS

1. Positive control is best described as *"What is intended to happen is what happens, and that is all that happens."*
2. Most of the time, safety is conceived in the mind of the performer before work starts by knowing what is to be accomplished and what to avoid.
3. The combination of technical expertise, chronic unease, conversations, and expert intuition of frontline workers provides the best source of safety for the unexpected.
4. FT is dangerous at CRITICAL STEPS and for their respective RIAs. Slow down to speed up.
5. **Hu** Tools exercise one or more aspects of RISK-BASED THINKING (anticipate, monitor, respond, learn), which is ST.
6. Entering the "line of fire" is a CRITICAL STEP.
7. Error traps in the workplace increase the likelihood of losing control. Unwanted variation in behavior leads to variation in results.
8. Landmines are "accidents waiting to happen." Landmines are hidden CRITICAL STEPS, unbeknownst to the performer, where a hazard is *poised* to trigger harm to an asset.
9. CRITICAL STEPS are *always* critical regardless of the presence of a hazard!

CHECKS FOR UNDERSTANDING

1. The RU-SAFE prework discussion guidance can be used during:
   a. Work preparation by frontline workers
   b. Work planning

    c. Work execution

    d. All the above

2. True or False. Positive control is absolutely necessary for RIAs as well as CRITICAL STEPS.

3. Fill in the blank. _____ thinking is necessary for the performance of CRITICAL STEPS.

4. If there is doubt about a CRITICAL STEP:

    a. Stop and then proceed when you feel confident.

    b. Stop any transfers of energy, movements of matter, or transmissions of information and get help from a competent individual.

    c. With the aid of a qualified co-worker or supervisor, verify the preconditions necessary for the CRITICAL STEP, and when convinced those conditions exist, proceed.

    d. b and c.

5. True or False. Frequent conversations—dialogue—among co-workers improve the understanding of the true state of technical processes.

(See Appendix 3 for answers.)

THINGS YOU CAN DO TOMORROW

1. Ask the top performers of a workgroup to describe techniques they adopted to help them perform their work properly and reliably. Contrast these techniques with the performance of the rest of the workgroup. Consider adopting/tailoring these techniques as **Hu** Tools for the organization's high-risk work.

2. With the aid and involvement of the workgroup, tailor the **Hu** Tools described in this chapter to the work of the organization.

3. Observe prework discussions and the respective field work, especially the high-risk portions. Identify adjustments workers make to accomplish the work. Understand the difference between the plan (*work-as-imagined*) and the actual behaviors (*work-as-done*), especially from a systems perspective.

4. Ask frontline personnel to choose/develop **Hu** Tools that aid positive control of CRITICAL STEPS. Consider the use of learning teams (see Glossary) for this purpose.

5. During a field observation, look for and reinforce the practice of stopping work when uncertain, emphasizing the need to slow down during high-risk situations.

6. Using Table 5.2, Common Workplace Error Traps, conduct a review of the presence of error traps during high-risk operations. Survey the organization's frontline workers to identify the most prevalent error traps.

7. Conduct a careful inspection of the workplace to assess the presence of landmines.

8. During event analyses, capture any losses of control, landmines, or failed CRITICAL STEPS. Identify and correct related organizational misalignments to support positive control of CRITICAL STEPS, adaptive capacity of frontline workers, and the ability to fail safely for future work.

REFERENCES

1 Howlett, H. (1995). *The Industrial Operator's Handbook: A Systematic Approach to Industrial Operations.* Pocatello: Techstar (p. 74).

2 Adapted from Institute of Nuclear Power Operations (1987, December). 'Hang in There.' *Lifted Leads,* 4(2).

3 Hollnagel, E. (2004). *Barriers and Accident Prevention.* Burlington: Ashgate (pp. 76, 121)..

4 *Adapted from National Rifle Association Gun Safety Rules.* Retrieved from: http://training. nra.org/nra-gun-safety-rules.aspx.

5 Hollnagel, E. (2009). *The ETTO Principle: Efficiency-Thoroughness Trade-Off: Why Things That Go Right Sometimes Go Wrong.* Burlington: Ashgate (p. 20).

6 Hollnagel, E. (2004). *Barriers and Accident Prevention.* Aldershot: Ashgate (p. 80). And: Trost, W., and Nertney, R. (1995). *Barrier Analysis* (DOE-01-TRAC-29-95). Idaho Falls, ID: Technical Research and Analysis Center, Scientech, Inc. (p. 10).

7 Hollnagel, E. (2009). *The ETTO Principle: Efficiency-Thoroughness Trade-Off: Why Things That Go Right Sometimes Go Wrong.* Burlington: Ashgate (pp. 25–30).

8 Gladwell, M. (2005). *Blink: The Power of Thinking Without Thinking.* New York: Little, Brown and Company (pp. 11–12).

9 Ibid. (p. 15). And: Kahneman, D. (2011). *Thinking, Fast and Slow.* New York: Farrar (pp. 415–418).

10 Salas, E., Rosen, M., and Diazgranados, D. (2009, October 28). 'Expertise-Based Intuition and Decision Making in Organizations.' *Journal of Management,* 36(4): 941–973. https://doi. org/10.1177/0149206350084.

11 Kahneman, D. (2011). *Thinking, Fast and Slow.* New York: Farrar (pp. 58, 416).

12 Muschara, T. (2018). *Risk-Based Thinking: Managing the Uncertainty of Human Error in Operations.* New York: Routledge (p. 244).

13 Ibid. (p. 117).

14 Gordenker, A. (2008, October 21). 'Pointing and Calling.' *The Japan Times.* Retrieved from: http://code7700.com/pointing_and_calling.htm.

15 Shinohara, K., Naito, H., Matsui, Y., and Hikono, M. (2013). 'The Effects of "Finger Pointing and Calling" on Cognitive Control Processes in the Task-switching Paradigm, International.' *Journal of Industrial Ergonomics,* 43: 129–136.

16 Muschara, T. (2018). *Risk-Based Thinking: Managing the Uncertainty of Human Error in Operations.* New York: Routledge (pp. 247–248).

17 During a personal conversation with Dr. James Reason in 1998, he said the chance for error in such "knowledge-based" performance situations is "a toss-up, 1 in 2," adding, "if you're lucky."

18 Plaschke, B. (1999, January 28). 'Looking for the Ultimate Tail-Gator?' Article in *LA Times.* Retrieved from: http://articles.latimes.com/1999/jan/28/sports/sp-2582.

19 Muschara, T. (2018). *Risk-Based Thinking: Managing the Uncertainty of Human Error in Operations.* New York: Routledge (pp. 48–52).

20 Swain, A., and Guttmann, H. (1983). *Handbook of Human Reliability Analysis with Emphasis on Nuclear Power Plant Applications: Final Report* (NUREG/CR-1278). Washington, DC: U.S. Nuclear Regulatory Commission.

21 Thomas, M. (2004). *Error Management Training: Defining Best Practices.* ATSB Aviation Safety Research Grant Scheme Project 2004/0050 (p. 13).

22 Hollnagel, E. (2014). *Safety-I and Safety-II: The Past and Future of Safety Management.* Farnham: Ashgate (p. 132).

23 Adapted from Institute of Nuclear Power Operations (1987, December). 'Half a Change is Worse than None.' *Lifted Leads,* 3(4).

6 Managing Critical Steps

Managers, ask for information you need to know, not what you want to hear. Workers, tell them what they need to hear, not what you want to tell them.

—Roger Boisjoly Former chief engineer for Morton-Thiokol, Inc.*

But performance does not mean "success every time." Performance is rather a "batting average." It will, indeed it must, have room for mistakes and even for failures. What performance has no room for is complacency and low standards.[1]

—Dr. Peter Drucker Author, educator, and management consultant

---

MARCEL LEDBETTER MOVING COMPANY[†,2]

Poor old Marcel, he's always tried to get in several kinds of businesses where he could make a profit and he wouldn't have to work so hard hauling pulpwood. He went into the moving business one time and got him a partner named James Lewis. Marcel Ledbetter Moving Company. He borrowed him some money, got him a few trucks, and one day the phone rang. "Mr. Ledbetter, will you move a piano for me?"

"Yes, ma'am!"

They got to the house, and it was a three-story house with a big bay window on the second floor. She wanted them to move the piano out of there and down to the ground. Marcel got up there and got to checking, and it wasn't no way. After the piano was done moved up there, they'd fixed the door some way. Marcel couldn't wedge the piano down, and he didn't have enough folks to tote it.

Marcel said, "I know what I'll do."

He got him one of them two-by-sixes and went there and nailed it on top of the house, stuck it out over the house, put him a block and tackle up there—one of them pulleys—brought the end down and into the bay window. Then he tied it around the piano, real good.

---

* Morton-Thiokol was the manufacturer of the solid rocket boosters used to launch the Space Shuttle Challenger in January 1986. Leaking O-rings due to the freezing weather triggered the loss of the shuttle and seven astronauts.

† This fictional story by the late comedian Jerry Clower, usually told orally and exuberantly in his southern conversational style before large audiences, is reprinted with permission of University Press of Mississippi.

DOI: 10.1201/9781003220213-6

James Lewis and the other hand went up there and was going to ease it out the window, and Marcel done wrapped the rope around his wrist down there on the sidewalk.

"All right, now, y'all be careful," yelling from the sidewalk below. "Shove it out easy, and I'm gonna ease it down."

They eased it out the window, and just as it left the ledge, that thing started down and Marcel started up with the rope tied around his arm. He passed that piano about halfway up, and the piano hit the sidewalk—boom!—went into a thousand pieces. Splinters covered the whole street. Marcel's head hit that pulley up there. Boom! Down he come flat on his back, right down on all that busted piano. Knocked him unconscious.

Here comes James Lewis down the steps. He got down and he slapped Marcel. "Oh, speak to me!"

Marcel opened his eyes. He said, "Why should I speak to you? I just passed you twice up there and you didn't say nothing."

By this point, it should be clear what the CRITICAL STEP is—*easing* the piano out the window. As soon as the piano left the windowsill, gravity took over (a sudden movement of solid matter—the piano). Marcell was not heavy enough to suspend the piano above the ground. Apparently, Marcell did not install enough "mechanical advantage" into the pulley (block and tackle) on the roof.* An option would have been to wrap the rope several times around a nearby tree, creating enough friction to support the piano's weight.

Two RIAs, mentioned in the preceding paragraph, if performed properly, would have created the necessary *condition* to allow Marcel to maintain positive control of the piano after it left the windowsill. However, Marcel and his helpers did not possess the technical expertise to tackle such a job. Regardless, how would you have *managed* this job, the CRITICAL STEP of lowering the piano out the bay window? What could have been done *before* the work began to have ensured success of the move; *during* the work to retain positive control; and *after* the work (incident) to improve their ability to plan and cope with future surprises?

Managing anything means being able to detect and observe it, to know where it needs to be, to determine if it's progressing properly; and being able to introduce corrective action effectively. RISK-BASED THINKING—anticipate, monitor, respond, and learn—supports the management function. Together, these logical reasoning processes are necessary in managing CRITICAL STEPS.

---

* A "block and tackle" is an assembly of a rope and pulleys that is used to lift loads. At least two pulleys are assembled to form the blocks, one that is fixed and one that moves with the load. The rope is threaded through the pulleys to provide *mechanical advantage* that amplifies that force applied to the rope. Source: www.constructionknowledge.net/general_technical_knowledge/general_tech_basic_six_simple_machines.php#3.

OBJECTIVES OF MANAGING CRITICAL STEPS

Recall that the principal goal of managing CRITICAL STEPS is to maximize the *success* of people in their everyday work—*creating value without losing control* of built-in hazards. We want to make sure the right things go right the first time, every time. This goal is accomplished by pursuing the following objectives:

1. Identify known (and recognize unknown) CRITICAL STEPS and their preconditions established by their respective RIAs.
2. Exercise positive control of the release of built-in hazards during the work.
3. Fail safely if a loss of control occurs during a CRITICAL STEP.
4. Align and realign the organization's system to support the preceding objectives.

Fundamentally, managing CRITICAL STEPS is a risk management process. It's not a human performance problem, *per se*. The goal is not so much focused on preventing human error as it is on making the right things go right—to be successful. This involves identifying, watching, controlling, and learning.[3] The practices introduced in this chapter address one or more of these objectives. As you can see, these objectives correlate closely with basic management: *plan, do, check, adjust*; and RISK-BASED THINKING: *anticipate, monitor, respond, learn*.

A CRITICAL STEP MENTAL MODEL

Dr. Erik Hollnagel developed a modeling method known as FRAM (Functional Resonance Analysis Method) to better understand the complexity and interconnectedness of how work is done.[4] Although seemingly complicated, we believe it's straightforward and helpful in thinking about—to analyze—a pending CRITICAL STEP. We adapted FRAM, illustrated in Figure 6.1, to help you think more systematically about CRITICAL STEPS and their performance—a mental model of CRITICAL STEPS. Whenever practicable, it's important to think systematically about CRITICAL STEPS. Obviously, *known* CRITICAL STEPS, those perpetually present whenever a particular operation is performed, must be analyzed in detail during engineering design, procedure development, and work planning, and reviewed during prework discussions. During the preparation stage of work, Figure 6.1 can assist engineers, planners, procedure writers, and even frontline workers to mentally simulate the performance and the effects of various conditions relevant to a proposed CRITICAL STEP. It's conducive to "what if" analysis.

However, in the workplace, frontline workers occasionally encounter landmines, surprise CRITICAL STEPS, previously unknown. If they are working remotely or do not have access to ready assistance, frontline workers could use the FRAM model to help them think and prepare themselves to perform an apparent CRITICAL STEP, if they must. A simplified FRAM can be applied on the go, to help workers plan and adjust one's response—the *think* function of STAR (self-checking)—to become more mindful of the work at hand, its goals, resources, expected outcomes, preconditions

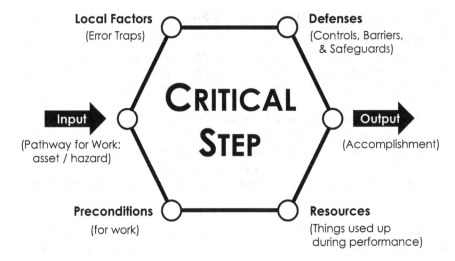

FIGURE 6.1 Mental model of CRITICAL STEPS. Hollnagel's FRAM analysis method is adapted to "model" and analyze a CRITICAL STEP, whether rigorously during CRITICAL STEP MAPPING or on the job. This illustration is repeated in Chapter 7, CRITICAL STEP MAPPING, for ease of reference.

Source: Adapted from figure 5.4 of Hollnagel, E. (2012). *FRAM: The Functional Resonance Analysis Method*. Boca Raton: CRC Press, pp. 46–53).

for safety, what to monitor, potential pitfalls, controls, contingencies, etc. The analysis method has utility during postwork reviews as well as being a tool to evaluate a CRITICAL STEP. Figure 6.1 illustrates the core elements of the analysis method, followed with a description of each element.

- *CRITICAL STEP*—The proposed human action that will trigger immediate, irreversible, and intolerable harm to an asset if that action or a preceding action (RIA) was performed improperly.
- *Input*—A cue to perform a CRITICAL STEP, usually signaled by a procedure step that creates a *pathway* for work or by the discovery of a *landmine*—changes in equipment condition, controls of hazards, or exposures of assets to hazards; output(s) of previous process/procedure steps or other human actions.
- *Output*—A change in the state of an asset; the intended value-added accomplishment (or harm, if performed improperly) of the CRITICAL STEP.
- *Preconditions*—Technical conditions, including pathways that must exist before the CRITICAL STEP is performed for the work; RIAs create some or all these conditions.
- *Resources*—These are things consumed or used up during performance of the CRITICAL STEP.
- *Local Factors*—Workplace conditions established by the work system that influence the performer's thinking and actions before and during the

CRITICAL STEP; that promote positive control; and *error traps* that negatively influence human performance, enhancing the potential for a loss of control during the CRITICAL STEP.

- *Defenses*—Controls, barriers, and safeguards needed for positive control and/or protection of assets (to fail safely, if possible); some RIAs establish defenses.

Additionally, the analysis method can be used to conduct a more detailed simulation of a loss of control of specific CRITICAL STEPS as described in Chapter 7.

THE PRACTICES OF MANAGING CRITICAL STEPS

Managers who implement change effectively in their organizations follow an iterative cycle toward accomplishing an initiative's goals: 1) plan, 2) do, 3) check, and 4) adjust. In its essence, *managing is about learning*. To manage anything involves basic gap analysis:

1. Identify the gap between where you are and where you want to be (goal).
2. Develop a strategy to close the gap (plan).
3. Implement the strategy (do).
4. Monitor progress of the plan in closing the gap (check).
5. Modify the plan to continue progress toward the goal (adjust).

Steps 4 and 5 obviously involve learning—checking and adjusting—while steps 1, 2, and 3 require prior learning. Without learning, management is ineffective. Managing CRITICAL STEPS attempts to control both known and unknown risks in operations, which demand ongoing checking and adjusting, and there are various occasions and ways of managing them—operational hazard control. Therefore, managing CRITICAL STEPS is referred to as a practice—you never stop learning. The practices subsequently described function collectively toward the perpetual operational goal of managing CRITICAL STEPS, executed through its four objectives mentioned earlier.

To manage CRITICAL STEPS systematically, it helps to recall the *Work Execution Process*, described in Chapter 3. The process is illustrated again in Figure 6.2. This adaptation of the process emphasizes four organizational functions: 1) engineering design, 2) work planning, 3) procedure development, and 4) the corrective/preventive action process, all important for managing CRITICAL STEPS.

The *Work Execution Process* suggests that three opportunities exist to manage CRITICAL STEPS: 1) before starting work, 2) during work, and 3) after completion of work. Before work is performed, CRITICAL STEPS can be identified during the engineering design phase of technical systems and during work planning/procedure development. Critical STEPS are necessary to complete work that is not allocated to machines. Therefore, work planners and procedure writers must carefully sequence and control the work to enable positive control of known CRITICAL STEPS that frontline workers must perform and fail safely if they lose control.

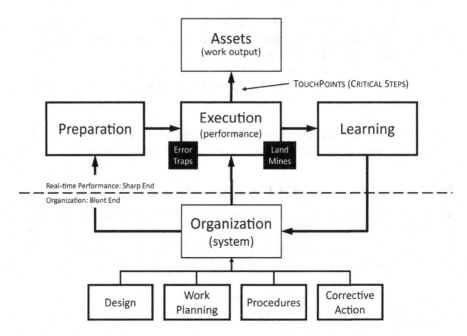

FIGURE 6.2 The *Work Execution Process* emphasizing the organizational functions involved in identifying and controlling CRITICAL STEPS. Feedback from workers is management's most important and richest source of information for SYSTEMS LEARNING.

The means or tools in managing CRITICAL STEPS operative at various stages of work are described in the following. More specific methods to identify and control CRITICAL STEPS may be effective for different technologies and industries.

Opportunities before Starting Work

1. *Engineering design*—With a foundation in user-centered design, inherently safer design, and system safety engineering, devise facilities and equipment with only the CRITICAL STEPS necessary to accomplish the organization's work. Minimize complexity, ambiguity, uncertainty, and volatility, among other error traps around their performance.
2. *Procedure development*—Identify and denote perpetual CRITICAL STEPS (and conditional CRITICAL STEPS) in current technical guidance and related procedures.
3. *Work planning*—Pinpoint CRITICAL STEPS, related RIAs, relevant error traps, and effective means of positive control for unique/one-off work not guided by approved technical procedures.
4. *Prework discussion*—Conduct a meeting with assigned personnel to discuss what is to be accomplished and what is to be avoided, including means of managing related risks, for a specific work activity.

Opportunities during Work

5. *Perform CRITICAL STEPS*—Exercise positive control of hazards in accomplishing work, being mindful of pathways, error traps, and landmines (hidden CRITICAL STEPS).
6. *Field observation and feedback*—Watch everyday work, looking for organizational factors that influence performance and safety, including good and at-risk practices.

Opportunities after Completion of Work

7. *Postwork review and reporting*—Pinpoint significant differences between *work-as-done* and *work-as-imagined* (planned).
8. *CA/PA process*—Realign the organization's system to strengthen the safety and reliability of operations using an institutionalized learning process.

Although you may aspire to establish *complete* control over the occurrences of CRITICAL STEPS in your operations, it is unrealistic to expect that to happen over the long term. Therefore, it's important to build in the ability to adapt to surprises and otherwise unforeseen opportunities and challenges. We end the chapter addressing this need—augmenting adaptive capacity.

ENGINEERING DESIGN

Although the scope of this book is limited to human performance within as-built, fixed facilities, modifications of existing equipment and systems and the introduction of new technologies are ongoing. It will seldom be necessary to modify existing facilities to manage CRITICAL STEPS. Occasionally, it will be. This practice focuses on the first three objectives of managing CRITICAL STEPS.

Good engineering adheres to the following order of precedence regarding human performance design: 1) prevent a loss of control (human error), 2) reduce the likelihood of losing control, 3) enable detection and correction of error and its outcomes (before the onset of harm), and 4) mitigate harm (after the onset of harm).[5] How? Engineers ensure their designs 1) avoid or minimize the occurrence of unnecessary CRITICAL STEPS through *inherently safer design* methods,* 2) optimize the operator's ability to maintain positive control and avoid dangerous error traps and landmines, adhering to *human-centered design* principles,[6] and 3) enable/preserve the adaptive capacities of the performer (described later in this chapter). Finally, the engineering design function blends engineering and safety from design inception.

---

* *Inherently Safer Design* methods create/modify technical processes that eliminate hazards, rather than accepting them and developing "add-on" functions to control/contain them. Eliminating hazards eliminates CRITICAL STEPS. According to the traditional hierarchy of controls, this is the most effective means of preventing events.

System safety—safety by design—determines the presence of safety by understanding the context of interactions between system components (humans, machines, and environment) and related built-in hazards.[7]

Despite the design principles just mentioned, no design would be complete without input from the intended users. Engineers and their managers must appreciate the risks they create for operators, builders, maintainers, and other end-users. However, their challenge is that they are ever more remote from work, in time and location, disconnected (organizationally insulated) from the everyday realities in which their designs will function.[8] Most engineers do not take educational courses on safety while in college or attend safety training after graduation as an employee. Therefore, designers must interact with frontline workers, safety professionals, and managers regularly, talking about human performance, operational risks, and system interactions encountered during high-hazard work.

So, what summary guidance can we offer engineers regarding CRITICAL STEPS and human performance? Dr. James Reason, in his book *Organizational Accidents Revisited*, suggests that engineers ask several questions about their proposed designs[9]:

- How easy is it for the user to understand what the [equipment] does? (Avoid adding complexity.)
- Is the range of possible actions readily apparent to the user? Are these features directly visible? (Avoid ambiguity and uncertainty.)
- Is it immediately obvious to users what they need to do to achieve a particular goal? (Avoid ambiguity.)
- Is it easy for them to perform the actions? (Avoid complexity.)
- Having performed the action, how easy is it for the user to find out whether the [asset, component, or system] is in the desired state? Is the state of the system readily apparent? (Avoid uncertainty.)

Similarly, Emery Roe and Paul Schulman, in their book *High Reliability Management*, strongly suggest that engineering design proposals and modifications should do the following[10]:

- Reduce volatility, uncertainty, complexity, and ambiguity (VUCA) (or not increase it) of the work faced by users.
- Preserve options (or not decrease flexibility) in responding to VUCA.
- Improve predictability and repeatability (or not diminish it).
- Minimize user's mental workload (or not add to existing).
- Add redundancies in barriers and safeguards along with backup resources in sustaining positive control and responding to off-normal/emergency situations (preserve tractability).

As an organization gains experience with CRITICAL STEPS, it will likely realize the occasional need for engineering solutions that will involve retrofitting equipment. Although engineering design cannot eliminate all CRITICAL STEPS, it can help manage (despite the characteristically slow response of engineering processes) the risk associated with the ones needed for work and to respond to the ones that creep into the technology and its operations.

PROCEDURE DEVELOPMENT

A written procedure specifies a sequence of human actions (instructions) for accomplishing a given objective—work—while protecting or limiting harm to the organization's assets. Therefore, procedure writers and process engineers *must* be mindful of the occurrence of CRITICAL STEPS in work plans. Procedures aid in the identification and control of known CRITICAL STEPS, denoting their presence in the operation. In most cases, they are denoted as *Cautions*, *Warnings*, or *Dangers*. It would be appropriate to insert a hold point before a CRITICAL STEP to verify the required preconditions exist—established by preceding RIAs—to perform the CRITICAL STEP safely.

> **Caution**: Be alert to the potential presence of *conditional* CRITICAL STEPS, instructions that follow an "if-this-then-that" logic. There may be occasions that the flow of work doesn't follow the norm but must take an alternative path. Also, these conditional paths may be hidden in computer-based procedures.

Additionally, procedure writers could help frontline workers more reliably exercise positive control by either eliminating or minimizing error traps at specific CRITICAL STEPS and related RIAs, such as haste, changes from routine, or violations of population stereotypes (up = off vs. up = on). An algorithm for systematically identifying and controlling perpetual (known) CRITICAL STEPS in existing technical procedures is described in Chapter 7.

For simple, low-hazard activities or work that involves no CRITICAL STEPS, a worker's knowledge and skill may be sufficiently reliable to accomplish the task without detailed checklists or formal procedures, allowing a greater degree of flexibility and discretion—and FT. In such situations, losing control anywhere in the task would not pose a serious threat to assets. Regardless, landmines abound, and workers must remain uneasy for any substantial work they do.

WORK PLANNING

Some work activities do not have standing technical procedures or standard operating procedures (SOPs) that otherwise guide recurrent work. Some work involves temporary, unique, or one-off situations, such as troubleshooting and corrective maintenance. If such work is complex or high risk, it must be planned to guide the performer in doing the work correctly, safely, and efficiently. Similar to procedures, work planning aids in identifying CRITICAL STEPS as well as promoting their positive control.

The first step for a work planner is to visit the work site and scope the work to define what is to be accomplished and its criteria for success. Conversations with frontline workers are encouraged. Second, the work planner develops the draft work instruction that specifies the series of tasks, denoting standards of performance (success criteria) to abide by. Many of the tasks performed will be regarded as *skill-of-the-craft* that do not require a checklist or procedure to perform. Some tasks in the work plan may have their own SOP, all of which may have hidden CRITICAL STEPS embedded in them. Be careful!

Like standing procedures, work instructions must highlight CRITICAL STEPS along with their respective preconditions (referencing RIAs if needed), again denoted as *Cautions, Warnings,* or *Dangers.* Work instructions for complex, high-risk work should be reviewed independently not only for technical accuracy but also for safety and reliability, which would involve an evaluation of CRITICAL STEPS (consider using the FRAM model illustrated in Figure 6.1) to validate identified controls and potential means of failing safely. The review is enhanced with a second physical walkdown of the work instruction *in the workplace.* This provides supervisors and key workers with a firsthand understanding of the work's demands, challenges, hazards, and pitfalls that a remote prework discussion cannot provide. The physical walkdown helps supervisors and workers identify CRITICAL STEPS and related RIAs not previously identified by work planners or standing technical procedures.

Planning includes scheduling. Work schedules must accommodate time for worker preparation, not just time for doing the work. Frontline workers and their supervisors need dedicated time to think about their work. This admonition goes for postwork reviews as well.

PREWORK DISCUSSION

In Chapter 3, we described prework discussions in detail, introducing **RU-SAFE**. But here we simply want to emphasize the workers' opportunity to understand the work's key assets and their hazards, to identify known CRITICAL STEPS, and how to exercise positive control of them as well as how to fail safely at each CRITICAL STEP.

PERFORM CRITICAL STEPS

Again, we must redirect the reader to Chapter 5 for details related to the in-field performance of CRITICAL STEPS and their related RIAs, especially the use of **Hu** Tools. We want to reinforce the cautious positive control of known CRITICAL STEPS and to sustain a posture of chronic unease regarding the possible presence of landmines—surprise CRITICAL STEPS. It is likely that landmines exist in every complex workplace. Therefore, frontline workers must be ready to respond conservatively when hidden or surprise CRITICAL STEPS arise. This also means that when there is a conflict between production goals and an asset's safety, subsequent actions are taken to protect assets. No one needs permission to protect assets.

FIELD OBSERVATION AND FEEDBACK

Too many people believe safety exists because frontline workers simply follow procedures—*work-as-imagined* by designers, procedure writers, and work planners. But there's no such thing as a perfect procedure. Because of the complexity of the workplace and even the work itself, people continually adjust to emerging workplace conditions to get work done. Applying the principles and practices of CRITICAL STEPS proactively addresses what must go right; "playing to win" so to speak, rather than preventing doing things wrong—"playing not to lose."[11]

Wouldn't it be worthwhile to know how normal, successful work is really performed? For managers, observation of work is perhaps the richest source of information about how their organizations support performance—learning. To understand why things normally go right, line managers must go into the workplace, watch everyday work, especially associated with activities that pose substantial risk to people, product, and property. When line managers watch work in real time, they see, firsthand, how frontline workers make things go right. To learn effectively, managers must personally interact with frontline personnel, asking follow-up questions about what was done (not in a judgmental way). Managers who don't understand the frontline workers' points of view in the workplace can miss the demands and constraints those workers regularly encounter.[12] While in the field, enlightened managers and supervisors take the opportunity to encourage the exercise of RISK-BASED THINKING and chronic unease, especially risks associated with CRITICAL STEPS. Although it is still important to find vulnerabilities in the workplace as well as with the conduct of work, it's always better to consider frontline workers as agents of the company rather than as adversaries. Relationships are important. (See the description of H&OP principle number one in Appendix 2: *People have dignity and inherent value as human beings.*)

Regular time spent on the shop floor is time well spent. But this time must be supported by the organization, aligned to achieve its learning purpose. Depending on the tempo of operations, managers should devote 1–4 hours a week watching work. Preferably, managers should show their face every day, asking questions, looking for ways to help. Giving line managers time to watch work helps them 1) appreciate "normal" work and the conditions people work in, 2) better understand behavior choices and how work goes right, and 3) realize how their systems enable or inhibit not only safety, but also quality, reliability, and productivity. The foregoing purposes cannot reasonably be accomplished in 5–10 minutes. Five minutes in the field reveals little about what is really happening—what we refer to as a "drive-by" observation.

Face-to-face interactions are individually focused and respectful. Feedback is a *two-way* process. Some call this interaction "engagement," where two people learn; learning occurs not only for the worker, but also for the manager. Face-to-face conversations between line managers and frontline personnel provide the most effective venue for understanding actual work conditions and performance and communicating standards, expectations, priorities, and values.[13] Ideally, the practical use of a manager's time in the field can be organized around three purposes:

1. *Reinforcing feedback*—The manager communicates with frontline personnel in the context of successful ongoing work, seeking to understand what is going right, reinforcing behavior choices that meet expectations, with a willingness to adjust expectations as the organization learns. This kind of feedback builds trust. Expect to devote most of the time here.

2. *Soliciting feedback*—The manager seeks input from frontline personnel on the value of work documents, tools, training, work conditions, etc., in the field, surprises encountered, and the effectiveness of related organizational support. This practice models the kind of conversations important to resilience. Understanding work from a systems perspective will enable managers to better understand how their organization is either aligned or misaligned to produce the desired work and business outcomes. The time devoted to this type of feedback is secondary to reinforcing feedback.

3. *Coaching and correcting feedback*—The manager gives frontline workers feedback about performance that does not meet expectations (such as at-risk choices), clarifying expectations and standards, demonstrating and guiding them on desired performance with the aid of work documents, training, and other performance resources. Reckless and unethical choices must be stopped by correcting the individual, while at-risk choices are coached to help the individuals understand the true risks and proper responses. Expect to devote the least amount of time here. However, this does not suggest it's unimportant.

To be most effective, managers should observe no less than the prework discussion and the performance of CRITICAL STEPS and related RIAs. Effective field observations include the following practices:

1. Provide feedback to frontline workers and their supervisors on:
   a. Desired behaviors (both proper adherence and conservative choices) associated with high-risk activities, especially actions that protect assets from hazards
   b. Behaviors that are either at-risk, ineffective, or improperly performed
   c. Unwanted behavior choices that are reckless or otherwise unethical
2. Receive feedback on system-related weaknesses and vulnerabilities that inhibit the safe and reliable conduct of work.
3. Encourage ongoing and robust conversations between frontline personnel and with support personnel if available.
4. Behavioral choices that warrant STOP-work or pauses in work (timeouts) provide real-time opportunities for identifying landmines, system weaknesses, and other latent conditions.
5. Encourage frontline personnel to reflect on their performance, using questions, soliciting their perspectives instead of simply telling them what you observed.
6. Debrief personnel in a quiet place, private if possible, to minimize distractions. Have a conversation about how to better handle the human performance risk, emphasizing the value of behavior choices to safety.

As suggested, it is helpful to have a set of questions that facilitate engaging the workforce prepared in advance to reference while observing work. Be careful not to distract workers during high-risk phases of the work. Some general questions you can tailor to your work group to promote conversations with frontline workers are as follows.

Example questions to ask yourself (as an observer):

- Is the work environment helping or hindering?
- How effective is supervision?
- Are expectations and procedures clear and usable, especially for high-risk operations?
- What can I do to ensure that "what must go right" indeed goes right?
- What operational aspects are working well? How can these be sustained?
- How are competent workers affirmed and encouraged? Is there an opportunity to reinforce them and use them to mentor/guide others?
- Should systems be revised to correct or sustain the work as done versus how it was planned?
- How can I facilitate those doing the work toward better outcomes, safer processes, or enhanced accommodation of their needs?
- Are my expectations enabling/inhibiting success?
- Are any system/organizational factors inhibiting safe and reliable performance?

Example questions to ask the worker (the performer):

- Is work being performed as anticipated? If not, how is work being performed?
- What aspect of the operation surprised you? Why?
- Where/when is the next event going to happen—what operations frighten you?
- Do you have the right tools, and are they useable? If not, what do you need?
- What adjustments are you making to ensure things go right?
- What workarounds have you encountered? Are they a long-term condition or temporary?
- Are procedures current, understandable, available, and used?
- Are you pleased with the training that prepared you to do this work?
- What communications or conversations can improve our operations?
- Are there opportunities to tailor or simplify the work to better support you?

Observations of work in the workplace provide great insights into the conduct of work. But there is another opportunity for managers after the work is done to gain insights about how the organization is aligned or not aligned to support safe and reliable work.

POSTWORK REVIEW AND REPORTING

Postwork reviews discuss and document significant differences between work preparation (*work-as-imagined*) and work execution (*work-as-done*). Conversations focus on why things went well, especially with surprise CRITICAL STEPS, and the related error traps or landmines. Workers must be given the opportunity after work's completion to provide feedback to their managers, so that they can act on it, reducing the time at risk for future work. Postwork reviews—learning after doing—collect fresh insights into the weaknesses and opportunities related to the technology, built-in defenses, processes, etc. The prime focus of the conversation is the organization and related management systems.

A postwork review is a brief meeting conducted after the completion of work that allows time for reflection and discussion, and for the collection of feedback from frontline workers and their supervisors. This meeting need not last more than 15–30 minutes. Research has reported that as little as 15 minutes devoted to reflection each day improved performance by more than 20 percent.[14] More effective postwork reviews provide a break after completion of work (in addition to a bathroom break and cleanup) to allow time for personal reflection before the meeting.

Recall that the simplified FRAM analysis method described earlier (Figure 6.1) can be used to evaluate adjustments made at CRITICAL STEPS. Such feedback should be communicated to line management, evaluated, and resolutions tracked to completion using the organization's CA/PA process (see next subsection). Remember, departures from *work-as-planned* should not always be perceived as negative. Despite weaknesses with procedures, conscientious workers attempt to improve both safety and efficiency. These behavioral choices just might be the most efficient and/or safest way to perform the task, so take note and determine the "why" before dismissing it.

To avoid inhibiting open conversations during the meeting, rank, seniority, and even experience should be temporarily set aside—everyone has an equal voice. Information collected in a postwork review is isolated from individuals' performance reviews—no one's career should be affected by what is said in a postwork review. We've said it before, and it's worth repeating. *It's not who's right, it's what's right!*

If your organization does not have a structure to guide a postwork review, the following protocol may serve as a springboard for developing your own approach. It is a modified version of what former fighter pilot James Murphy (author of *Flawless Execution* (2005)) used to guide combat mission debriefs in the U.S. Air Force[15]:

1. *What did we intend to do?*—Review the prework discussion and other preparation activities (what to *accomplish* and what to *avoid*—*work-as-imagined*).
2. *What did we actually do?*—Identify what worked well (pluses) and what did not go well (deltas), including surprises, especially around CRITICAL STEPS (*work-as-done*). It's also important to note as-left conditions. Ask people to tell their stories of what happened, goal conflicts and their resolutions, and adjustments made to achieve their goals.
3. *What was different and why?*—Compare what actually happened to what was planned. Briefly explore the reasons for the differences (deltas).
4. *Report it*—Inform supervision or management of the important pluses and deltas, including suggestions, using approved communication methods to ensure follow-up.

A more rigorous approach to a postwork review, originated in the U.S. Army, is known as an after-action review.[16] Remember, if a job is important enough to have a prework discussion, it is important enough to have a postwork review, always if the work included CRITICAL STEPS.

We believe frontline workers have a moral duty to report significant issues encountered, highlighting the adjustments made, even when nothing bad happened. If not reported, someone or something could get hurt or damaged in the future. Silence is a problem multiplier.[17] Line managers would do well to actively and regularly solicit such information from their workforce and reward those who do. Frontline workers may not recognize the importance of such differences from a safety perspective once *work-as-done* becomes routine. Reports of persistent differences between *work-as-done* and *work-as-imagined* (drift) offer managers insights into organizational misalignments that may adversely influence frontline personnel's ability to exercise positive control of CRITICAL STEPS and/or to adapt to surprises or losses of control.

CORRECTIVE ACTION/PREVENTIVE ACTION PROCESS

Learning doesn't occur until behavior changes—improvements in both the individual and the system. When there are significant and/or persistent differences between *work-as-done* and *work-as-imagined*, the system is misaligned.* Improvement (system realignment) depends on an *understanding* of problems from a systems perspective, including the review of local and organizational factors that contribute to the effectiveness of defenses and behavior choices. For learning to be effective in the long term, line managers must appreciate the need for follow-up conversations with those who report an issue to understand its context. Therefore, managers and CA/PA support staff must understand systems thinking (see Figure 8.1). It is beyond the scope of this book to describe systems thinking thoroughly; but an introductory primer is provided in the section "Execution Requires Systems Thinking," in Chapter 8.

Effective corrective actions will be more organizational and systemic in nature,[18] which tend to improve defenses against harm, reducing the severity of events that still occur. However, you must avoid the temptation to adopt corrective/preventive actions† that are local remedies rather than system-related solutions. People-focused corrective actions tend to drive down frequency but have negligible impact on severity. System solutions tend to be costlier and require more managerial effort than local remedies, which place more of the burden on frontline personnel to adapt to system-generated flaws. It is line management's responsibility to align/realign the organization and its systems to support safe, reliable, and resilient operations.

A formal, institutionally approved CA/PA process is necessary to encourage the development and implementation of effective corrective/preventive actions as well as promote managerial accountability for follow-through with their implementation.

---

* Alignment refers to the extent to which the system's various management systems and organizational functions work together to achieve its business mission and goals, which includes safety, quality, and reliability, as well as productivity and profitability. See Muschara, T. (2018). *Risk-Based Thinking: Managing the Uncertainty of Human Error in Operations.* New York: Routledge (pp. 58–60).
† Corrective actions are organizational responses to an event to contain the harm done and to prevent recurrence of the event, while preventive actions are proactive solutions to identified risks of events that have yet happened.

Effective SYSTEMS LEARNING involves implementation and integration of corrective or preventive actions that realign the organization to support and sustain productive and resilient performance. Occasionally there will be no actions in response to an event. Only actions that warrant careful implementation and follow-up are managed with a formal CA/PA process. Less important actions can be managed using approaches deemed appropriate by the manager. Briefly, an effective CA/PA process possesses the following characteristics:

- A structured communication method provides the members of the organization with a means of reporting situations to management, documenting events, close calls, deficiencies, nonconformances, and potentially dangerous conditions.
- Corrective/preventive action development takes advantage of a diversity of technical and organizational perspectives with an eye on the system and less on the individual.
- Long-term corrective/preventive actions do not rely solely on disciplinary action or admonishments of frontline workers to simply "pay attention" or "follow procedures."
- Managers are wary of simple procedure revisions as a default corrective/preventive action. More rules do not necessarily equate to more safety.
- Training as a corrective/preventive action is used only when knowledge or skill deficiencies hinder the performance of CRITICAL STEPS or RIAs. Comparatively, training is expensive.
- Line managers are accountable for the timely and effective implementation of corrective/preventive actions, especially those associated with CRITICAL STEPS of high-risk operations.
- Postponement of the implementation of corrective/preventive actions for high-risk activities, prolonging the time at risk, is strictly avoided except for extraordinary circumstances. Organizations and their managers don't tolerate unnecessary CRITICAL STEPS and/or faulty or missing defenses at known CRITICAL STEPS.
- Effectiveness reviews of corrective/preventive actions are regularly conducted to ascertain if they have been implemented, have been integrated into organizational processes in a way that will prevent recurrence, and have indeed resolved the related issue.

SUMMARY—PLAN | DO | CHECK | ADJUST

Effective management is resilience in action—that is, anticipate, monitor, respond, and learn. Effective managers readily adopt and integrate the practices of RISK-BASED THINKING because of its similarity to the functions associated with the management cycle. Table 6.1 summarizes these overlapping practices in managing CRITICAL STEPS during the *Work Execution Process*.

TABLE 6.1
Summary of the Management Practices Relevant to Managing CRITICAL STEPS.

Management Cycle	Plan	Do	Check	Adjust
Management outcome	*Work-as-Imagined*	*Work-as-Done*	Differences	Realignment
Work Execution Phase	Preparation (before work)	Execution (during work)	Execution/learning (during/after work)	
Means	• Engineering design • Procedure development • CRITICAL STEP MAPPING • Work planning • Prework discussion	• Perform CRITICAL STEPS	• Observation and feedback • Postwork reviews	• CA/PA process
Cornerstones of RISK-BASED THINKING	Anticipate	Monitor/Respond		Learn

Note: The means of exercising RISK-BASED THINKING are listed relative to the three phases of *Work Execution Process.*

AUGMENTING ADAPTIVE CAPACITY

Murphy's law is wrong! What could go wrong usually goes right. Despite complications, surprises, and insufficient resources, operations tend to proceed without failure mostly because of the expertise and alertness of frontline workers. People, though fallible, do much more than simply follow procedures—they think and adjust.[19] Conditions today are rarely the same as yesterday, and tomorrow will be different—in a VUCA work environment. Workers *create* safety as they sort out goal conflicts, closing gaps between plans and reality.[20] The system "works" most of the time, not despite, but *because* operators deviate from expectations to accommodate the flaws of work-as-imagined/designed, they become the system's buffers or shock absorbers.[21]

> Good operators rely extensively on knowledge-driven monitoring technical expertise instead of rote procedural compliance. This practice allows operators to detect problems before they become significant, to compensate for poor design of procedures, to distinguish instrumentation failures from component failures, and to become better aware (in a deep sense) of the unit's current state.[22]

Procedure adherence is still the primary means of operational hazard control, but with a mind poised to predict, detect, and respond to uncertainty and risk. No one can write a procedure that anticipates all the situations a user will encounter at a specific place and time. Nothing is perfect, and nothing is always as it seems in a complex workplace. Therefore, frontline workers need latitude for safety—*adaptive capacity*—in the workplace, able to respond with some degrees of freedom to unforeseen work situations—to *do* safety and achieve success. Adaptive capacity—making adjustments—bridges the gap between the way things are and what they need to be.

Remember, safety is NOT the absence of events. Safety is the *presence* of defenses-in-depth and flexible human actions in the workplace. To quote Dr. David Woods, "Safety is a verb."[23] Safety is what you DO to protect assets—controls, barriers, and safeguards. Adaptability in the workplace is an organization's most important safeguard.[24] To make things go right, frontline workers must 1) understand the "big picture" and know the key assets that are important in the current work activity, 2) recognize threats to those assets (pathways, touchpoints, error traps, and landmines), 3) realize when technical guidance for the situation is uncertain or ambiguous, 4) seek technical assistance from reliable sources, and 5) make conservative decisions—adjust—that sustain positive control and/or protect assets from potential harm . . . at CRITICAL STEPS.[25]

Adaptive capacity is a system property, and as a "property," it can and must be designed or built into (engineered) the organization's system to exercise positive control of known risks *and* to support adaptive responses to unknown risks. Adaptive capacity can be managed by aligning the organization's structure to promote technical expertise, teamwork, flexibility, RISK-BASED THINKING, slack, and situation awareness, collectively to add value, and avoid or minimize harm.[26] An understanding of the strengths and weaknesses of human beings helps in aligning these six organizational factors to support workers' adaptive capacity.[27] Without going into detail that is beyond the book's scope, Table 6.2 describes the aforementioned organizational factors and their purposes, and provides example means that enhance workers' chances for success, making appropriate adjustments, especially in an environment of volatility, uncertainty, complexity, and ambiguity.[28]

> **Caution**: Although adaptive capacity is necessary to adjust to ambiguities and inaccuracies in *work-as-imagined*, too much individual autonomy has a downside risk. Local fixes can hide underlying organizational weaknesses.[29] Significant and recurring field adjustments MUST be reported to management to allow them to make adjustments at a system level.

> **Caution**: Take care to preserve adaptive capacity during engineering-related modifications of facilities and equipment. This is true also for revisions to management systems and related administrative policies and practices.

After CSM and prework discussions, field work is the third possible occasion in which to identify CRITICAL STEPS—these pose the greatest risk to assets because they were unanticipated. Frontline workers with a persistent sense of unease—a mindfulness of pathways—readily predict and detect landmines in time to make adjustments or to stop the work and consult with others before proceeding headlong into a high-risk activity. Their expert intuition kicks in to alert them to danger. However, people should not be given complete freedom to do whatever they please. When in doubt, people should stop the work and get help before making adjustments on the fly. **If there is doubt, there is no doubt—STOP!** Then get help.

TABLE 6.2
Organizational Factors and Related Practical Means of Augmenting Adaptive Capacity for Frontline Workers in High-Hazard, Industrial Work.

Organizational Factors	Description* (Including Their Purposes)	Practical Means (Things You Can Do to Set People Up for Success)
Technical expertise	• The comprehensive knowledge, understanding, skills, experience, and proficiency relevant to the technology, assets, and built-in hazards • To serve as a technical backup to the facility's design and its procedures • To enhance expert intuition • To build management's confidence in the ability of frontline personnel to detect uncertainty and threats and to respond conservatively	• Technical training and qualification—initial, refresher, and skills practice (under stressful conditions); emphasis on fundamental knowledge—first principles; system mental models; contingencies (off-normal) and emergencies • Knowledge of key assets and their critical parameters (safe operating envelope (SOE)); knowledge of facility's/assets' safety design bases (safety analyses) • Integrated operations—awareness and understanding of operations upstream and downstream of individual's job/task • Proficiency—developing experience on the job and performing within some required periodicity (e.g., every 6 months); line-oriented flight training[30] • Review of industry and in-house operating experience (lessons learned) relevant to the work at hand • Postwork reviews—emphasizing what worked well (or not) and why/how; differences between work-as-done and work-as-imagined; reporting • Human performance training; "error wisdom" and "foresight training"[31]
Teamwork	• The ability of persons to collaborate in accomplishing a task that requires two or more people • To improve information flow among members and groups • To remove fear of speaking up, of making conservative choices • To be willing to sacrifice resources (trade-offs) for the safety goals of frontline operations • To enhance the accuracy of situation awareness	• Communication—open, robust conversations; redundant channels; three-part communication; humility: willingness and desire to give and receive feedback; centralized during wide-scale emergencies and disruptions • Just culture—treating people with dignity and respect; socially safe to allow people to disclose controversial opinions; not punishing individuals for one-off mistakes regardless of their severity • Networking—seeking diversity of opinion about risk from other persons or groups; willingness to sacrifice resources for the sharp end; looking out for each other • Teamwork training, e.g., crew resource management[32]

(Continued)

TABLE 6.2 (Continued)

Organizational Factors	Description* (Including Their Purposes)	Practical Means (Things You Can Do to Set People Up for Success)
		• Devil's advocate—skeptics; persons purposefully taking critical position relative to proposed solutions on an issue to provoke deeper thinking on it
		• Timeouts—regular updates for team members on system/work status during work stoppages; hold points for CRITICAL STEPS
		• Roles and responsibilities related to the business' reasons for the work: safety, quality, production, and reliability
Flexibility	• The capability of systems to adjust to new challenges; the provision of alternatives to accomplish an objective	• Redundancies—defenses-in-depth; duplicate/backup protections, e.g., skills, servers, routers, tools, parts, equipment, information, software, etc.
	• To solve problems without disruption, preserving options	• Organizational structure—preplanned real-time shifts of authority to decentralized control during peak demand and emergencies; clearly understood organizational values; deference to expertise[33]
	• To accommodate goal conflicts and avoid working at cross-purposes	
	• To provide frontline workers degrees of freedom to improvise and experiment when assets are not exposed to hazards—no CRITICAL STEPS in play	• Mode confusion (software)—provision for human/manual control of automatic systems, when system status is uncertain; make system status obvious to user
		• Bundling—consolidation of two or more functions into a single effort that accomplishes all; skill-of-the-craft; cross-trained frontline workers; multi-function tools
		• Autonomy—allowing exercise of individual judgment in the performance of low-risk work; STOP-work authority and execution of contingencies to preserve safety
		• Procedure use—follow procedures as written unless compliance would lead to harm; avoid rote compliance as if following a recipe
RISK-BASED THINKING	• Logical (slow) thinking about risks to assets relative to their workplace hazards: anticipate, monitor, respond, and learn	• **Hu** Tools—workplace non-technical skills[34] that enable RISK-BASED THINKING practices: anticipate, monitor, respond, and learn
	• Sensitivity to operations, the place where work is done and danger exists	• Responding—taking necessary actions to exercise positive control and/or protecting assets
	• To promote chronic unease—a readiness for surprise, a preoccupation with failure	• Chronic unease—remind frontline staff of hazards and to respect them; continuous sensitivity to variability of Risk-Important Conditions and to abnormal and unusual situations; mindfulness of pathways for harm and impending transfers of energy (ΔE), movements of matter (ΔM), and transmissions of information (ΔI) relative to key assets
	• To be willing to subordinate production to safety when encountering goal conflicts	

Organizational Factors	Description* (Including Their Purposes)	Practical Means (Things You Can Do to Set People Up for Success)
	• To improve surety of positive control and means to fail safely	• Access to information—real-time availability to technical, system, and organizational information; communication channels; available system diagrams and flow charts; procedures and checklists; intuitive menus (software), etc.[35] • Priming—foreknowledge of what to accomplish/avoid; prework discussions • Skepticism—wary of good news/acceptance of bad news; validate facts and assumptions; ask questions: "What if" scenario simulations; "Why not?"/walkthroughs/ tabletop exercises; "What can go wrong (and when it does, what will you do?)" • Field observation—line management monitoring of field operations; feedback that reinforces the practices of RISK-BASED THINKING: anticipate, monitor, respond, and learn and chronic unease; reinforce/ reward discovery, reporting, admission of mistakes • Procedures—technically accurate, usable, and available; available technical bases documentation
Slack[36]	• The margin (bandwidth and buffers) built into work relative to resources, time, and space • To respond to or cope with surprises in work environment—helps avoid overwhelming adaptive capacity • To acknowledge criteria for acceptable/successful performance	• Staffing levels—technical expertise bench strength; accommodate fluctuations in qualified employees • Time—margin in the schedule (time available exceeds the time required for the task) • Resources—tools, spare parts, information, ready reserves, and safety equipment; ability to repurpose and reallocate • Backups—communication channels; contingency plans • Space—physical space to maneuver • Workload management—match workload and tempo of operations; eliminate unnecessary requirements/workload; minimize the number of things operators must attend to • Workarounds—eliminate dependence on workers to accommodate long-term equipment deficiencies, especially at CRITICAL STEPS

(Continued)

TABLE 6.2 (Continued)

Organizational Factors	Description* (Including Their Purposes)	Practical Means (Things You Can Do to Set People Up for Success)
Situation awareness	• The perception of the elements in the working environment, comprehension of the situation, and potential impact on next steps • To develop and sustain an accurate mental picture of current operations • To know current configuration of operating equipment, systems, and threats • To recognize when the current plan no longer works or threatens safety	• Engagement—ongoing monitoring; mental involvement in automated activities to combat boredom and complacency • Alertness—health and fatigue management; workload management • Equipment/system status—configuration control; startup readiness checks; means of ascertaining the state of equipment or systems; actual versus expected Risk-Important Conditions; review of equipment status at shift turnover; digital equipment status board; equipment operating trend logs; facility-wide, public announcements/ alerts of major equipment changes • "Noise" reduction—remove distractions and minimize interruptions; housekeeping; eliminate unnecessary demands on memory and attention that distract from production, safety, and reliability[37] • Clarity—eliminate or reduce volatility, uncertainty, complexity, and ambiguity associated with technology and related work processes • **Hu** Tools—prework discussions, timeouts, hold points, callouts, turnover/handover • Environmental conditions—lighting, equipment layout; component labeling, noise and vibration; habitability

* The description of each organizational factor should provide sufficient insight to provoke creativity and imagination to identify and/or develop other practical means not listed for that factor.

MANAGING HUMAN PERFORMANCE—FUNDAMENTALS

In practice, the integration and implementation of new practices involve a sustained focus on adherence to H&OP principles (especially by line managers) and mindful application of new practices (especially in the workplace). Expectations along these lines have proven to be essential for long-term success in any human endeavor. There can be no *real* accountability without clear expectations. We encourage the following approach[38]:

1. *Form expectations*—Identify what, where, how, when, and standards of new practices (to learn); what practices to stop (to unlearn). Obtain the agreement of target populations.
2. *Communicate expectations*—Conduct regular conversations on expectation, including why; conduct formal training (knowledge and skill development).

3. *Implement*—Go live after a break-in period. Align the system to enable new behaviors and/or to inhibit old practices to provide tools, resources, incentives/disincentives, and feedback.
4. *Check*—Solicit/provide feedback (reinforce, coach, correct) to/from the workforce. Identify improper organizational support (misalignments). Dedicate time for field observations. Actively solicit feedback from post-work reviews.
5. *Adjust*—Learn from successes and failures. Revise expectations if necessary. Realign the system to enable (support) desired practices/conditions or to inhibit undesired practices/conditions.

KEY TAKEAWAYS

1. The goal of managing CRITICAL STEPS is not so much focused on preventing human error as it is on making sure the right things go right, first time, every time, while avoiding or limiting harm to important assets.
2. The goal of managing CRITICAL STEPS is accomplished by pursuing the following objectives:
   a. Identify known (and recognize unknown) CRITICAL STEPS and their preconditions established by their respective RIAs.
   b. Exercise positive control of the release of built-in hazards during the work.
   c. Fail safely if a loss of control occurs during a CRITICAL STEP.
   d. Align and realign the organization's system to support the preceding objectives.
3. Managing CRITICAL STEPS is a risk management process: plan, do, check, adjust. Managing, if it is to be effective, requires aggressive learning.
4. Identifying and controlling CRITICAL STEPS and their related RIAs are key outcomes of RISK-BASED THINKING: 1) anticipating, 2) monitoring, 3) responding, and 4) learning.
5. CRITICAL STEPS can be modelled, which allows them to be managed during all phases of work.
6. The *Work Execution Process* suggests that three opportunities exist to identify and establish positive control of CRITICAL STEPS: 1) before starting work, 2) during work, and 3) after completion of work.
7. The following organizational functions and field activities provide the means of managing CRITICAL STEPS (to accomplish its goal and four objectives):
   a. Engineering design
   b. Procedure development
   c. Work planning
   d. Prework discussion
   e. Perform CRITICAL STEPS
   f. Field observation and feedback
   g. Postwork review and reporting
   h. CA/PA process

8. Building adaptive capacity into the organization helps frontline workers create safety as they resolve goal conflicts, closing gaps between plans and reality. Workers "finish the design."
9. The adoption of new behaviors or the cessation of old practices must be managed intentionally and at an organizational level (to align/realign).

CHECKS FOR UNDERSTANDING

1. True or False. Frontline workers should have the freedom to diverge from approved procedures whenever they feel it is expedient for accomplishing work goals.
2. The goal for managing CRITICAL STEPS is:
   a. Eliminate all CRITICAL STEPS
   b. Prevent human error
   c. Accomplish business goals regardless of cost
   d. Identify and control CRITICAL STEPS
   e. Improve reliability
3. The purpose of field observation by managers is to:
   a. Assess the activity's results against expected results
   b. See firsthand the difference between *work-as-done* and *work-as-imagined*
   c. Understand how the system supports or inhibits performance
   d. Foster conversation about risk and what must go right
   e. Provide constructive feedback to workers
   f. All of the above
4. True or False. Managing is learning. Why/why not?
5. Frontline worker adaptive capacity improves the organization's _____ to threats of harm to assets.

(See Appendix 3 for answers.)

THINGS YOU CAN DO TOMORROW

1. For a select organizational unit, prepare a list of technical procedures prioritized by risk and frequency of occurrence. Develop a workdown curve (plan) to systematically identify CRITICAL STEPS and related RIAs in each procedure. CSM should be applied in this process.
2. Develop expectations for managers to conduct regular field observations of work activities, specifying the criteria for a good one. Ensure their weekly schedules afford sufficient time to conduct them (avoiding drive-by observations).
3. Benchmark the organization's CA/PA against a known exemplar. Identify strengths, weaknesses, opportunities, and threats, especially around line management's accountability for its use and effectiveness.
4. At a venue where several design engineers are present, explore the CRITICAL STEP concept with them. Explore how the design process minimizes the occurrence of CRITICAL STEPS and establishes positive control of those that remain in a design.

5. Assess the quality of postwork reviews performed, including the alignment of the organization in support of them. Discuss the workforce's willingness to report problems and opportunities, especially surprise CRITICAL STEPS and significant differences between *work-as-done* and *work-as-imagined* discovered during work.

6. Assess the organization's support of the frontline workforce's adaptive capacity (see Table 6.1). Explore ways to strengthen their ability to adjust to surprises safely and reliably and to hone their expert intuition.

REFERENCES

1 Drucker, P. (1974). *Management: Tasks, Responsibilities, Practices.* New York: Harper and Row (p. 456).

2 Clower, J. (1992). *Stories from Home.* Jackson: University Press of Mississippi (pp. 11–112). Reproduced with permission.

3 Hollnagel, E. (2014). *Safety-I and Safety-II: The Past and Future of Safety Management.* Farnham: Ashgate (p. 121).

4 Hollnagel, E. (2012). *FRAM—the Functional Resonance Analysis Method: Modelling Complex Socio-technical Systems.* Farnham: Ashgate. And: Hollnagel, E. (2018). *Safety-II in Practice: Developing the Resilience Potentials.* New York: Routledge (pp. 114–120).

5 Sgobba, T. (ed. in chief), et al. (2018). *Space Safety and Human Performance.* Cambridge: Butterworth-Heinemann (pp. 556–561).

6 Norman, D. (2013). *The Design of Everyday Things* (revised and expanded ed.). New York: Basic (pp. 8–10).

7 Leveson, N. (2018). 'System Safety.' In: Sgobba, T. (ed. in chief), et al. (eds.). *Space Safety and Human Performance.* Cambridge: Butterworth-Heinemann (pp. 278–281).

8 Reason, J. (2016). *Organizational Accidents Revisited.* Burlington: Ashgate (p. 36).

9 Ibid. (pp. 38–39).

10 Roe, E., and Schulman, P. (2008). *High Reliability Management: Operations on the Edge.* Stanford: Stanford University Press (p. 213).

11 Hollnagel, E. (2014). *Safety-II in Practice: Developing the Resilience Potentials.* Abington: Routledge (p. 101).

12 Nemeth, C., et al. (2014). *Resilience Engineering in Practice, Volume 2: Becoming Resilient.* Burlington: Ashgate (p. xiv).

13 Hopkins, A., and Maslen, S. (2015). *Risky Rewards: How Company Bonuses Affect Safety.* Boca Raton: CRC Press (p. 114).

14 Gino, F., and Staats, B. (2015). 'Why Organizations Don't Learn.' *Harvard Business Review,* 93(11): 110–118.

15 Murphy, J. (2005). *Flawless Execution.* New York: Regan Books (pp. 133–160).

16 United States Army (2013). *The Leaders Guide to After Action Reviews.* Combined Arms Center—Training (CAC-T) Training Management Directorate (TMD) Fort Leavenworth, Kansas. Retrieved from: https://pinnacle-leaders.com/wp-content/uploads/2018/02/Leaders_Guide_to_AAR.pdf.

17 Quote attributed to Laurin Mooney during ORCHSE H&OP (virtual) Summit, January 12–14, 2021.

18 Perin, C. (2005). *Shouldering Risk: The Culture of Control in the Nuclear Power Industry.* Princeton: Princeton University Press (pp. 208–209).

19 Woods, D., Dekker, S., Cook, R., Johannesen, L., and Sarter, N. (2010). *Behind Human Error* (2nd ed.). Farnham, UK: Ashgate (p. 8).

20 Ibid.
21 Perin, C. (2005). *Shouldering Risk: The Culture of Control in the Nuclear Power Industry.* Princeton: Princeton University Press (pp. 208–209).
22 Mumaw, et al. (2000). 'There is More to Monitoring a Nuclear Power Plant than Meets the Eye.' *Journal of the Human Factors and Ergonomics Society*: 51–53.
23 Presentation by Dr. Woods on March 14, 2019, entitled "Resilience is a Verb." Video Retrieved from: www.youtube.com/watch?reload=9&v=V2qj5gMsjrU&feature=emb_title.
24 Reason, J. (2008). *The Human Contribution.* Burlington: Ashgate (p. 239).
25 Watts-Englert, J., Woods, D., and Patterson, E. (2018). 'Resilient Anomaly Response in Mission Control Center.' In: Sgobba, T. (ed. in chief), et al. (eds.). *Space Safety and Human Performance.* Cambridge: Butterworth-Heinemann (pp. 581–582).
26 Woods, D. (2019). *4 Essentials of Resilience, Revisited.* Retrieved from: www.research-gate.net/publication/330116587_4_Essentials_of_resilience_revisited.
27 Schutte, P. (2018). 'Human-Machine Interaction.' In: Sgobba, T. (ed. in chief), et al. (eds.). *Space Safety and Human Performance.* Cambridge: Butterworth-Heinemann (pp. 454–456).
28 Roe, E., and Schulman, P. (2008). *High Reliability Management: Operating on the Edge.* Stanford: Stanford University Press (p. 121).
29 Reason, J. (2008). *The Human Contribution.* Burlington: Ashgate (pp. 258–259).
30 Harris, D. (2011). *Human Performance on the Flight Deck.* Farnham: Ashgate (pp. 262–264).
31 Reason, J. (2008). *The Human Contribution.* Burlington: Ashgate (pp. 246–251). Error wisdom training informs and instills the mental skills to cope with human performance risks in the workplace. Foresight training develops the mental skills to recognize the indications that something is not right.
32 Kanki, B. (ed. in chief), et al. (2010). *Crew Resource Management.* San Diego: Academic Press.
33 Roe, E., and Schulman, P. (2008). *High Reliability Management: Operating on the Edge.* Stanford: Stanford University Press (pp. 138–139, 232–233).
34 Flin, R., O'Connor, P., and Crichton, M. (2008). *Safety at the Sharp End: A Guide to Non-Technical Skills.* Burlington: Ashgate (pp. 1–13).
35 Schutte, P. (2018). 'Human-Machine Interaction.' In: Sgobba, T. (ed. in chief), et al. (eds.). *Space Safety and Human Performance.* Cambridge: Butterworth-Heinemann (pp. 451–453).
36 Patriarca, R. (2021, January 12). *Slack: A Key Enabler of Resilient Performance.* Retrieved from: www.resilience-engineering-association.org/blog/2021/01/12/slack-a-key-enabler-of-resilient-performance/.
37 Perin, C. (2005). *Shouldering Risks: The Culture of Control in the Nuclear Power Industry.* Princeton: Princeton University Press (pp. 211–213).
38 Connors, R., and Smith, T. (2009). *How Did That Happen? Holding People Accountable for Results the Positive, Principled Way.* New York: Penguin.

7 CRITICAL STEP MAPPING

A risk assessment is nothing more than a careful examination of what, in your work, could cause harm, so that you can weigh up whether you have taken enough precautions or should do more to prevent harm.[1]

—U.K. Health and Safety Executive

The map is not the territory.[2]

—Dr. Alfred Korzybski Independent scholar

COSTLY CLERICAL ERROR[3]

Israel Corporation, Ltd. is a large holding company that deals mostly in fertilizers and specialty chemicals. On what looked like a normal trading day, within minutes the company's share price reportedly dropped from $46,385 down to $58.

Investors on the Tel Aviv Stock Exchange (TASE) reacted nervously, setting off a wave of electronic selling. In just over 5 minutes, the single trade shrunk the Tel Aviv 25, an index of the top 25 traded companies, by 2.5 percent. This dramatic change triggered rules that halted trading automatically. Officials began an investigation immediately. The reason for the temporary crash: a typing error by a clerk. The trade was cancelled when the mistake was recognized. In fact, officials cancelled all trades made within that 5-minute window and the market recovered.

The momentary plunge in the TASE was traced to the one sale order. A clerk intended to type in the symbol of another, lower-priced company, but instead typed in the trading symbol for the Israel Corp Ltd. (ILCO). Such mistakes are commonly referred to as a "fat finger" error.[4] We don't know the symbol of the other company, but it's likely that it was similar to ILCO. The clerk, perhaps in a hurry or due to habit, mindlessly depressed the Enter key without checking. We don't know for sure. The fact that one stroke of a key could trigger such a dramatic plunge of value illustrates the dangers of operating in today's digitally driven, complex markets.[5]

By now we hope you recognize the CRITICAL STEP. The pathway for loss was created when the ILCO symbol was typed into the "sell" field of the digital form. Typing trading symbols into digital forms is normal and repetitive. The all-important touchpoint/CRITICAL STEP is the Enter key, which when depressed triggered the loss. That's when the information was transmitted to signal a sell of ILCO shares, albeit unintended—a trading error. Do you think

DOI: 10.1201/9781003220213-7

the Israel Corporation made any corrective actions in how clerks prevent trading errors? Don't know. It might be worthwhile to require a peer-check for high-value trades—checking the symbols, number of shares, and values typed into a digital form *before* depressing the Enter key. Automation doesn't correct this kind of problem; a human being still must enter this information somewhere, sometime.

WHAT IS CRITICAL STEP MAPPING?

CSM is a table-top analysis process for identifying those human actions, procedure steps, or phases of a work activity that will cause serious harm—death, injury, damage, or loss—to one or more of an organization's key assets should the performer lose control of work. Obviously, you should consider CSM for those work activities that involve intrinsically high-hazard operations. Perhaps, the Israel Corporation would have detected and corrected the aforementioned vulnerability in their trading process before suffering such an event had they conducted a similar analysis.

CSM is more like hunting (searching/exploring) than following a map. *You are looking for human actions embedded in a prescribed work activity that match the definition of a CRITICAL STEP.* CSM reveals specific human actions that could breach the safety boundaries for the assets involved in a process or work activity. Once denoted or recognized, a CRITICAL STEP alerts the performer to the "edge of the cliff" (a pathway), which is usually defined explicitly by critical parameters that define the asset's limits of safety. These are the points or stages in the work process that need positive control and defenses to keep from falling off the cliff or catching one on the way down, figuratively.

CSM is preparatory in nature, not part of the job. It is a form of failure modes and effects analysis (FMEA),[6] but without the emphasis on identifying every human error or hazard. Instead, it helps identify where an asset's protection from harm is most important—what must go right to be successful. This approach methodically identifies assets, hazards, touchpoints, CRITICAL STEPS, their related RIAs, and defenses for a specific task, operation, process, or procedure. If, during the analysis, you realize an asset will undergo a significant change in state, you've arrived at a CRITICAL STEP.

MAPS AND THEIR LIMITATIONS

Maps are used to navigate between points on the map, negotiating the most efficient and safe passage between those points. The destination is the value you desire to create, but there are hazards en route. To arrive safely, you'll have to navigate the hazards represented on the map. Good maps closely represent reality, the so-called lay of the land. Procedures are similar to maps. They describe the technical and physical conditions of the workplace before the activity begins as well as pinpointing its goal—the value or end state to be achieved. However, be forewarned: procedures, just like maps, were prepared some time ago, and both maps and procedures become less accurate over time. Things change. There is no such thing as a perfect map. The map is not the territory. Similarly, there is no such thing as a perfect procedure. *The procedure is not the work—work-as-done is never the same as work-as-imagined!*

Caution: As stated, the map is not the territory. The procedure is a starting point. CSM attempts to identify CRITICAL STEPS in *work-as-imagined* as prescribed in procedures based on conditions assumed by the procedure's authors and their managers. On the day of the work, reality is possibly different from the initial conditions assumed in procedures.

WHEN CAN CSM BE USED?

CSM can be used when it's realized that people will perform high-hazard work activities, working in direct contact with hazardous processes. It's a work-planning tool, not a tool to be used in a prework discussion. The CSM method is generic and can be applied in any work domain that involves worker control of high-hazard operations. Occasions to use CSM may include the following:

- *System design or modifications*—decisions regarding the allocation of functions to people that cannot be fulfilled by automation or physical systems; building in defenses and enabling adaptive features to enhance positive control and failing safely; validation of human factors design at known CRITICAL STEPS
- *Procedure development*—sequencing activities to accomplish one or more business objectives, denoting perpetual CRITICAL STEPS and highlighting them with appropriate warnings and cautions
- *Work planning*—developing guidance for one-off or infrequently performed work that has no standing, preapproved procedures; checks the existence of positive control and the ability to fail safely

CSM is a relatively lengthy, table-top process that should not be used during prework discussions on the day of the work. CSM is conducted before the work is scheduled to identify perpetual CRITICAL STEPS and select necessary defenses. By the time the work is scheduled, procedures should already denote known CRITICAL STEPS.

TEN-STEP CRITICAL STEP MAPPING PROCESS

The CSM process (denoted in Figure 7.1) involves three phases: 1) preparatory, 2) mapping, and 3) protection (not to be confused with the *Work Execution Process*). The preparatory phase (steps 1 through 3) identifies the work and its goals, assets, and hazards; and estimates the severity of potential harm. The mapping phase (steps 4 through 7) distinguishes between human actions, touchpoints, CRITICAL STEPS, and related RIAs. Once these are identified, then the protection phase (steps 8 through 10) helps analysts identify and evaluate the defenses (controls, barriers, and safeguards) needed to avoid a loss of control,

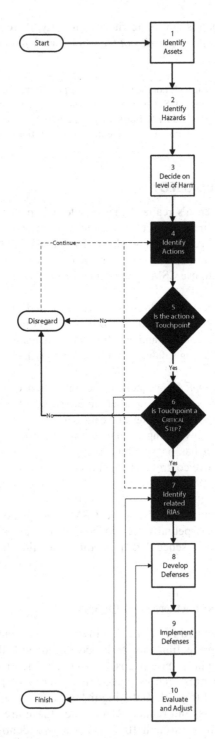

FIGURE 7.1 Ten-step CRITICAL STEP MAPPING analysis process.

protect related assets, or mitigate harm after the onset of harm (to fail safely, if practicable).

Note: For management purposes, it may be advisable to develop a risk-based work-down schedule for all high-risk work activities, starting with the most frequent, highest-risk procedures.

Prior to the first use of the process, analysts should feel free to revise the wording and phrases of the process to reflect the language of the particular work domain or organization. CSM is best done by process engineers and procedure writers with the aid of experienced frontline staff. But before starting CSM, it is important to define the scope of the analysis. Typically, work processes will have a variety of phases or stages. The work can be broken out into its basic processes or segments, and analysis can be done for such segments and does not have to be performed for the entire work activity at one sitting.

To prepare for the analysis, we suggest reviewing and posting the definitions of key terms used in the analysis: human action, touchpoint, CRITICAL STEP, RIA, controls, barriers, and safeguards. Everyone should have the procedure in hand, whether hardcopy or digital, while the range of steps under scrutiny is projected onto a screen in the meeting room to allow everyone to see that set of steps concurrently. If an approved technical procedure is unavailable, we suggest developing a *functional flow block diagram* (FFBD) of the work before conducting this analysis.* Analysts should use the current revision of the approved procedure or work guidance, assuming the existence of initial conditions specified in the procedure. Alternatively, it may be helpful to conduct a field walkdown of the work activity before starting, to give support staff a clearer understanding of the context of the work. The mapping process entails the following ten steps:

Note: The number of assets determines the scope of the mapping project.

1. *Identify assets*—Intent: Know the assets important to the organization relevant to the work at hand. These key assets must be protected from harm.
   a. Define the job's purpose and its business goal(s)—i.e., accomplishments (work outputs that remain after job completion)—which involve the use of and/or changes to a key asset(s).
   b. Ask: What key assets contribute to the success of this task, considering the following business purposes?

- Safety	- Reliability	- Others?
- Productivity	- Quality	

   c. Know the criteria for success for each work output.

---

* An FFBD is a simple block diagram of the work functions to be accomplished for activities not guided by an approved procedure. It is about *what* must be done, NOT about who, why, when, where, or how. Keep it simple. All functions start with a verb, a human action, for example, start pump motor, open valve, read patient barcode, adjust temperature.

d. Consider creating an "asset register" for each task or operation, denoting the asset's critical parameters.[7] This asset register could be expanded to record perpetual CRITICAL STEPS and suggested defenses (similar to a risk register).*

2. *Identify hazards*—Intent: Know the built-in hazards (sources of energy, matter, and information) required to create value but that would threaten the safety of assets during work.

   Note: It is helpful to review the procedure's precautions and limitations and to conduct a field walk-through to improve the chances of identifying hazards to assets.

   a. Identify the key functions—the work to be performed during the activity.
   b. Pinpoint built-in hazards necessary to achieve work outputs.

3. *Decide on the severity of harm*—Intent: Define the degree of harm (injury/damage/loss) that would be considered *intolerable* and irreversible. Management should make this decision with technical input from responsible engineers. Not all harm is "critical." Some level of harm to a key asset may be tolerable and recoverable with minimal cost.

   a. Review previous events (internal and external to the organization) and worker experiences associated with the work, relevant to the assets.
   b. Characterize the impact (not probability) of the harm if control is lost. Reference the organization's event classification criteria to types of events—degree of 1) injury to people, 2) costs (opportunity and recovery), 3) interruptions/delays, 4) loss of mission functionality, 5) property/environmental damage, 6) security breaches, 7) loss of quality, etc.
   c. Test: Would there be an *event* if the asset suffered the degree of damage, loss, or injury characterized previously? Would it be necessary to identify corrective and preventive actions?
   d. The degree of harm described here will define *intolerable harm* for a CRITICAL STEP in Step 6. If keeping an asset register, record this information there for reference during the mapping phase.

   Note: This concludes the preparatory phase of CSM. The following step begins the mapping phase.

   Caution: Experience has revealed that confusion occurs when analysts attempt to consider multiple assets for each procedure step. It is preferable to complete a pass through the mapping and protection phases of the algorithm considering one asset at a time.

4. *Pinpoint human actions*—Intent: Identify human actions in the operation, task, or work. All human actions involve a force applied over a

---

* A risk register is a document used to log and track the management of risks, such as nature of the risk, reference, owner, and mitigation measures. Risks might be safety risks, commercial risks, financial risks, environmental risks, and more. It is typically created at the start of a project and is regularly referenced and updated throughout the life of the project.

distance—i.e., work—but may not involve work on an asset. The analysis focuses on the verbs used in the guiding document or the FFBD.

  a. Read each step for each operational function (step-by-step), looking for verbs that direct a human action.

  b. Decide if verbs are clearly defined and denoted in the procedure.

  c. Reference: Use a local procedure writer's guide for the definition of action verbs (if available).

Note: Frequently, this part of the analysis uncovers weaknesses with the procedure's wording and layout. Some verbs may have multiple meanings; or there may be multiple verbs for similar actions, such as check, ensure, and verify, which can have subtle differences in expectations for the performer. For sophisticated procedures, it helps to have a human factors specialist or qualified/experienced procedure writer present during the analysis to suggest appropriate improvements to the usability of the procedure.

5. *Is the action a touchpoint?* —Intent: Pinpoint which worker actions involve direct interaction with assets and hazards.

  a. Clarify the asset and the hazard for the human action.

  b. Identify pathway. Describe the means of work used: 1) transfers of energy (ΔE), 2) movements of matter (ΔM), or 3) transmissions of information (ΔI) that threaten the integrity of the asset. No pathway, no work.

  c. Ask: Would the human action do work on the asset, such that the state of the asset changes (regardless of whether the change in state is desired or otherwise)?

    i. No. Disregard the step. Proceed to the next procedure step or human action.

    ii. Yes. The human action is a touchpoint. Proceed to Step 6.

6. *Is a touchpoint a CRITICAL STEP?*—Intent: Identify CRITICAL STEPS from touchpoints.

  a. Ask: Does the touchpoint satisfy the definition of a CRITICAL STEP? (See Step 7 for a description of *improper* performance.) Verify that the harm of a loss of control of the hazard exceeds the severity criteria denoted in Step 3.

    i. No. Disregard and proceed to the next procedure step or human action.

    ii. Yes. The touchpoint is a CRITICAL STEP. Proceed to Step 7.

7. *Identify related RIAs*—Intent: Identify earlier human actions that establish the preconditions for the CRITICAL STEP.

  a. Ask: What Risk-Important Conditions must exist:

    i. To accomplish the production objective?

    ii. To perform the CRITICAL STEP safely?

    iii. To fail safely?

  b. Ask: What preceding human actions create one or more of the preconditions mentioned earlier?

Note: The following analysis is optional but may be helpful in validating the logic of Steps 6 and 7 and orienting the analyst for the protection phase of the analysis.

Using Figure 7.2, <u>run</u> various *mental simulations* of the CRITICAL STEP and RIAs being considered, preferably by experienced workers. Mental simulations help <u>characterize</u> what *will* happen to the asset if the CRITICAL STEP (or RIA) is performed *improperly*, which can be assessed using the following HAZOP* guidewords for action performed improperly at the specific touchpoint.[8] With "X" being the CRITICAL STEP, <u>ask</u> what if:

- No/Not X (omit X, proceeding to the next step without performing it)
- More/Less X (too much or not enough X)
- As well as Y (another action in addition to X)
- Reverse of X (wrong direction/opposite of X)
- Other than X (an action Y instead of X)
- Wrong object
- Fast/Slow X (too quickly or too slowly)
- Sooner than Y (precondition) (performing X, before Y is established)

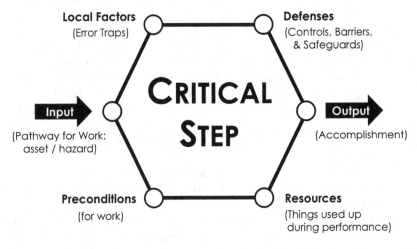

FIGURE 7.2   Analyzing CRITICAL STEPS. This illustration is a duplication of Figure 6.1 for ease of reference.

Source: Adapted from figure 5.4 of Hollnagel, E. (2012). *FRAM: The Functional Resonance Analysis Method*. Boca Raton: CRC Press (pp. 46–53).

Note: This step concludes the mapping phase of CSM. The following step starts the protection phase, still considering one asset at a time. As you consider how to exercise control of each CRITICAL STEP and means of failing safely, we encourage you to refer to Figure 7.2 frequently. Figure 7.2 is a duplication of Figure 6.1 in Chapter 6 for ease of reference, but that each element of CRITICAL STEP model is described in Chapter 6.

---

* HAZOP—Hazard and Operability—This is an analysis process that uses "guide words" to help analysts identify problems with an activity associated with deviations from design or procedural intent or losses of control.

8. *Develop defenses*—Intent: Using RISK-BASED THINKING (anticipate, monitor, respond, and learn), develop or confirm that controls, barriers, and safeguards exist for CRITICAL STEPS and related RIAs; for the protection of assets; and for the control of hazards.

   a. Ask frontline personnel for their suggestions on the selection and effectiveness of defenses.

   b. Match defenses with the risks: controls for CRITICAL STEPS, barriers to moderate ΔE, ΔM, or ΔI, and safeguards to mitigate harm to assets.*

   c. Annotate CRITICAL STEPS and related RIAs uniquely in procedures or work plans.

   d. Eliminate or control dominant error traps at each CRITICAL STEP and at related RIAs.

   e. Identify at-risk practices to avoid—actions that increase chances of a loss of control.

   f. Specify contingencies (what to do when ____ happens).

   g. Specify STOP criteria that signal the approach to an asset's safety limits (safety-critical parameters that define the asset's SOE).

   h. Enhance the frontline worker's adaptive capacity to respond to the unexpected: to avoid potential damage, to take advantage of opportunities, or to respond to unintended consequences.

9. *Implement defenses*—Intent: Install or verify defenses that promote positive control of human actions and protection of assets should a frontline worker lose control.

   a. Conduct a *local factors analysis*[9] for new behaviors, **Hu** Tools, etc., necessary to sustain the new behaviors.

   b. Verify the presence and operability of barriers and safeguards.

   c. Align the organization to support new behaviors, to eliminate support for unwanted behavior choices (at-risk practices), and to maintain barriers and safeguards.

10. *Evaluate and adjust*—Intent: Validate the effectiveness of defenses.

   a. Observe frontline worker performance periodically to validate adoption of and effective use of various **Hu** Tools and other new practices, especially at CRITICAL STEPS and RIAs.

   b. Ascertain the presence and robustness of barriers and safeguards that aid in protecting assets and failing safely.

   c. Resolve serious and persistent differences between *work-as-done* and *work-as-imagined*.

   d. Ask frontline workers for their opinions of the value and effectiveness of controls, barriers, and safeguards and whether additional modifications are needed.

---

* It is recommended that the reader study Dr. Erik Hollnagel's book on defenses, *Barriers and Accident Prevention* (2004), in particular pages 81–108.

ADVANTAGES AND DISADVANTAGES OF THE CSM PROCESS

ADVANTAGES

1. Identifies perpetual CRITICAL STEPS for a prescribed work activity or a new work activity
2. Provides a detailed understanding of each identified CRITICAL STEP and related RIAs
3. Provides insight into the operational needs of frontline personnel
4. Can be applied to any task in any industrial domain
5. Can be applied formally or informally
6. Takes advantage of the technical expertise of line personnel and fosters frontline ownership of the technical process
7. Requires minimal training; usually a one-time, one-day session to introduce the methodology
8. Does not require special tools or software, just a meeting room and a means of projection of the work guidance
9. Does not require analysts to be competent in human behavior, human factors methods, interviews, observations, or questionnaires
10. May identify weaknesses with equipment/system design, workplace layout, and procedure structure, usability, and wording
11. Promotes technical, interdisciplinary relationships that enhance risk-related conversations
12. Promotes the development and implementation of necessary defenses (controls, barriers, and safeguards)

DISADVANTAGES

1. May not identify CRITICAL STEPS unique to local conditions, such as other work in the area, physical location of the work, built-in sources of energy, landmines—that exist on the day and time the task is performed
2. May not identify CRITICAL STEPS embedded in skill-of-the-craft actions, use of high-powered tools, or other skills (chunks of actions for a particular task)
3. Can be time-consuming for large, complex work activities
4. Takes line personnel off the shop floor away from otherwise productive activities
5. Does not evaluate cognitive functions or human error (can be perceived as an advantage)
6. Requires the analysts to possess an in-depth technical understanding of the work activity or task

KEY TAKEAWAYS

1. CSM is a table-top analysis process for identifying those steps, actions, or phases of a work activity that <u>will</u> cause serious harm—death, injury,

damage, or loss—to one or more of an organization's key assets should the performer lose control of work.

2. CSM is a preparatory process, not part of prework discussions or the work. It can be used for new or existing work activities.

3. There is no such thing as a perfect procedure. The procedure is not the work—*work-as-done is never the same as work-as-imagined!*

4. CSM reveals perpetual CRITICAL STEPS; it does not identify CRITICAL STEPS that are unique to the task's context specific to those doing the work, the time and place of the work, and luck.

5. The mapping phase of CSM distinguishes between human actions, touchpoints, CRITICAL STEPS, and related RIAs.

6. The mapping phase should evaluate one asset at a time to identify CRITICAL STEPS. This way, the team will not get confused during the process as to which asset is being evaluated.

CHECKS FOR UNDERSTANDING

1. True or False. All CRITICAL STEPS are touchpoints, but not all touchpoints are CRITICAL STEPS.

2. Which of the following is the deciding factor in whether a loss of control at a touchpoint is a CRITICAL STEP?
   a. Consequence is irreversible.
   b. The harm resulting from the touchpoint is immediate.
   c. The severity of harm to an asset exceeds a predefined threshold.
   d. All of the above.

3. True or False. RIAs can be identified for an asset before the CRITICAL STEP is known.

4. True or False. All RIAs are touchpoints.

(See Appendix 3 for answers.)

THINGS YOU CAN DO TOMORROW

1. Using Figure 7.1 as a springboard, tailor the CSM process to accommodate your organization's language and methods. Consider using a local learning team to develop the organization's approach to identifying and controlling CRITICAL STEPS.

2. Review previous events with frontline workers in a small group to identify the relevant CRITICAL STEP that triggered the harm. Include a discussion of the RIAs and what could be done to establish positive control and protect assets.

3. Apply CSM for all new/modified technical/operational processes as part of your risk management process.

4. Incorporate CSM into maintenance testing control and operations lock-out/tagout processes, especially for the removal of an isolation. The sequence of isolation removal may contain one or more hidden CRITICAL

STEPS (landmines) during restoration of energy and piping systems to service.

5. Post the CSM flow chart and the definitions of a touchpoint, CRITICAL STEP, and RIA on the wall of the room in which CSM occurs. Consider posting these definitions anywhere that prework discussions occur.

REFERENCES

1 U.K. Health and Safety Executive (2003, July). '5 Steps to Risk Assessment.' *INDG163(rev1)*. Retrieved from: www.aberdeenshire.gov.uk/media/7334/market_risk_assess_guide.pdf.

2 Korzybski, A. (1994). *Science and Sanity: An Introduction to Non-Aristotelian Systems and General Semantics* (5th ed.; originally published in 1933). Englewood: Institute of General Semantics (p. xvii).

3 Lev, D. (2013, August 25). 'Clerk's Error Causes TASE to Tank.' *Arutz Sheva*. Retrieved from: www.israelnationalnews.com/News/News.aspx/171263.

4 Manuel, D. (2013). *The Israel Corp Ltd Drops Nearly 100% After "Fat Finger" Error*. Retrieved from: www.davemanuel.com/2013/08/25/tel-aviv-stock-exchange-hit-by-typo/.

5 Berman, J. (2013, August 26). *Israel Corporation's Share Price Drops 99.8 Percent Because of Typo*. Retrieved from: www.huffpost.com/entry/israel-corporation-share-price-drops_n_3816480.

6 Stamatis, D. (2003). *Failure Mode and Effect Analysis: FMEA From Theory to Execution*. Milwaukee: American Society for Quality, Quality Press.

7 Viner, D. (2015). *Occupational Risk Control: Predicting and Preventing the Unwanted*. Farnham: Gower (pp. 79–80).

8 Kirwan, B., and Ainsworth, L. (1992). *A Guide to Task Analysis*. London: Taylor and Francis (pp. 194–201).

9 Muschara, T. (2018). *Risk-Based Thinking: Managing the Uncertainty of Human Error*. New York: Routledge (pp. 170–172, 201–203).

8 Integrating and Implementing
CRITICAL STEPS

If you want to change the way you think, change the words you use.[1]

—**Dr. Karl Weick Author:** *Managing the Unexpected* **and Professor Emeritus**

We don't think ourselves into a new way of acting, we act ourselves into a new way of thinking.[2]

—**Larry Bossidy Author:** *Execution: The Discipline of Getting Things Done*

Initiatives fail most often not because the strategy is wrong but because they are not executed well.[3] *Execution*—getting things done and sustaining them—is based on basic management (plan, do, check, and adjust) and *follow-through*. Successful, sustained execution requires integration as well as implementation. *Integration* enables new ways of thinking and acting based on principles, while *implementation* institutionalizes programmatic requirements. Implementation involves basic management skills, while integration requires leadership commitment and engagement. Both are necessary to execute CRITICAL STEPS successfully for the long term.

Remember. The principal goal of managing CRITICAL STEPS is to maximize the *success* of people—those who create value for the company, frontline personnel in direct contact with the technology's built-in hazards. We are talking about success not only in the business domain of safety but also in the domains of quality, reliability, productivity, and even profitability. Managing CRITICAL STEPS applies to all fields of human performance,[4] and human performance risk exists in all domains of business. Therefore, identifying and controlling CRITICAL STEPS are strategic to the organization's success. Long-term success with managing CRITICAL STEPS depends on daily and precise implementation of the following objectives:

1. Identify known (and recognize unknown) CRITICAL STEPS and their preconditions established by their respective RIAs.
2. Exercise positive control of the release of built-in hazards during the work.
3. Fail safely if a loss of control occurs during a CRITICAL STEP.
4. Align and realign the organization's system to support the preceding objectives.

DOI: 10.1201/9781003220213-8

To establish and then sustain these risk-based practices, the organization must be aligned and continually realigned, undergirded by line management's commitment and follow-through, their understanding and application of systems thinking, and wisdom shaped by RISK-BASED THINKING.

The practice of managing CRITICAL STEPS—identifying and controlling them—should not be thought of as a quick solution to an organization's human performance challenges. This initiative is not a one-and-done affair—it's an ongoing way of managing human performance risk in the workplace. But to achieve and sustain execution of the principles and practices of managing CRITICAL STEPS, it's important to better understand the contrast between integration and implementation.

INTEGRATION VERSUS IMPLEMENTATION

The role of implementation is to develop and establish the practices of managing CRITICAL STEPS. A program—a set of requirements—can be developed to handle perpetual (known) CRITICAL STEPS in operations. That's an important first step in managing CRITICAL STEPS, and we believe this book provides the insights and recommendations to establish an effective management system.

Integration, however, is the *ongoing process of learning and unlearning*, of aligning and realigning the system to support the enduring accomplishment of the objectives listed earlier. Integration nurtures the frontline workforce's technical expertise, the will to communicate, and the capacity to adapt and respond to emergent conditions (unknown) in the workplace. The foregoing depends on their understanding and adoption of the principles of H&OP (Appendix 2). Eventually, these new ways of thinking, acting, and learning becomes the normal way of doing work.

Risk is an ever-present workplace dynamic, whether known or unknown, and must be managed by principle as well as by directive. Programs and structure help manage the known risks, while principles, technical expertise, and RISK-BASED THINKING, among others, help handle the unknown risks. Safety, reliability, quality, etc., are not accomplished simply by compliance with a set of rules and criteria—a program. The strong suggestion is that besides incorporating safety margins, redundancies, and directives into system design, organizational structure, and operation, safety is otherwise created and sustained by strengthening the adaptive capacity of frontline personnel. Frontline workers themselves become a safety margin for the designers of the work.[5] This kind of thinking cannot be developed and sustained by implementing a program alone but through a program augmented with the H&OP operating principles, RISK-BASED THINKING chronic unease, teamwork, and the development of expert intuition. This is the role of integration.

Be forewarned! When managing CRITICAL STEPS devolves into just another program, people stop thinking mindfully about risk. When managers believe (assume) they have completed all the requirements, they will think they're safe—a compliance mentality. The workforce soon follow suit. Unfortunately, when safety is treated as a set of requirements, frontline personnel then think this way too, following the example and edicts of their managers and supervisors mindlessly. Compliance-based thinking does not breed success over the long term.

EXECUTION REQUIRES SYSTEMS THINKING

Organizations are made up of people, groups of people, and its equipment doing things together—all the parts interacting and functioning to produce a desired business result. Systems thinking offers managers a way of understanding how their organization *really* works. Managers structure their organizations to produce a business result: *work-as-imagined.* But conditions change and the organization and its various components can interact and function in obscure ways, producing work practices and outcomes different than originally imagined, whether for good or ill: *work-as-done.* So that you do not endure such uncertainty when implementing CRITICAL STEPS and without going into detail, we briefly describe systems thinking to get you thinking rightly about your organization.

Systems thinking transcends conventional linear thinking: *x* causes *y,* an oversimplification of complex systems most of us are so used to. Alternatively, systems thinking, as a more perceptive framework for organizational thinking, considers the connections between people and the technology, people and management systems, people and the environment, and people and people—the conditions people work in. Systems thinking explains the differences between *work-as-imagined* and *work-as-done.* Cause-and-effect thinking is still useful in understanding people's behavior choices but applied in a more contextual way.

Work-as-done will almost always differ from *work-as-imagined,*[6] which suggests the presence of misalignments between organizational factors and behavior choices in the workplace. This framework of influences is illustrated in Figure 8.1.

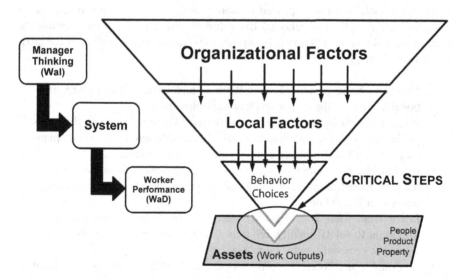

FIGURE 8.1 Systems thinking for H&OP. Systems thinking helps explain the difference between *work-as-imagined* (WaI) and *work-as-done* (WaD). For the sake of simplicity, the feedback channels necessary for systems to function are not shown.

Source: Muschara, T. (2018). *Risk-Based Thinking* (p. 127).

Systems are not perfect—procedures, technology, policies, resources, training, etc., are always underspecified. People's choices are influenced by multiple local working conditions (local factors), which are, in turn, influenced by various organizational factors, such as its management structures, training programs, engineering functions, human resources, values, and priorities. Influences imply interactions, relationships, and feedback. People, regardless of where they work in the organization, are part of the system; and the system must aid, not inhibit, desired behaviors and outcomes—this is the intent of alignment and realignment. Remember: realignment—SYSTEMS LEARNING—depends on feedback, which is not shown in Figure 8.1 (for simplicity).

> **Caution**: Figure 8.1 is a simplification of reality; social and technical organizations are more complex than illustrated. As a model of how systems thinking works, it is an incomplete description of all the aspects involved in day-to-day operations. But the framework is useful in helping managers *think* more deeply about how their organizations work, prompting them to ask better questions and make better decisions.

ATTRIBUTES AND PRINCIPLES THAT STRENGTHEN INTEGRATION

Studies in resilience engineering reveal that strategic approaches to managing H&OP and related challenges of complex operations exhibit some common attributes. These attributes, when exhibited at both the individual and organizational levels, have contributed to improved reliability and greater capacity to respond to opportunities as well as to rebound from threats and setbacks. Implementing the management of CRITICAL STEPS is strengthened by the integration of following attributes (or precepts)[7]:

- *Expect to be surprised*—Belief: Work is underspecified at an organizational level and subject to workplace variables and luck.
- *Possess a chronic sense of unease*—Beliefs: There are no such things as perfect procedures, complete training, flawless design, or foolproof planning; and past performance is not a guarantee of future success.
- *Be capable of bending (not breaking)*—Beliefs: Risk is dynamic. Safety is the presence of defenses-in-depth and adaptive capacity, able to minimize harm after losing control.
- *Learn from what goes right as well as from failure*—Belief: Lessons important to safety, quality, reliability, and productivity are embedded in work that goes well—what people do.
- *Embrace operational humility*—Beliefs: Frontline workers are local experts. No one can know all there is to know.
- *Acknowledge differences between work-as-imagined and work-as-done*—Beliefs: Surprise happens, and safety is what the worker does to protect assets. Differences persist as long as misalignments exist between the organization and the workplace.

- *Value collaboration and welcome different points of view*—Belief: Managing variety requires variety and diversity of input.
- *Manage operations using systems thinking*—Belief: Performance emerges from complex, interconnected, and interdependent systems and their components.

In addition to these attributes, the building blocks of managing H&OP are founded on a set of principles (see Appendix 2) that shape thinking, decisions, and actions. Although H&OP is more an operating philosophy than a program, managing CRITICAL STEPS does require some programmatic direction to implement. Yet sustaining the practices—through integration—requires adherence to principles, which, unlike procedures, serve as a compass—core beliefs—in guiding responses to workplace risks.[8]

1. People have dignity and inherent value as human beings.
2. People are fallible.
3. People do not purposefully come to work to fail.
4. Behavior choices are predictable and manageable.
5. Risk is an inherent, dynamic feature in the way an organization operates.
6. Organizations are perfectly tuned to get the results they are getting.
7. Conditions that spawn tomorrow's events exist today.

> **Caution**: Long-term success of implementation depends on integrating the afore-mentioned attributes and principles that will guide the thinking and choices of an organization engaged in high-risk operations. Persistent deviations from these core beliefs will seriously hinder sustained effectiveness of managing CRITICAL STEPS.

A FOUR-STAGE APPROACH TO EXECUTING A CRITICAL STEPS INITIATIVE

Introducing a new way of performing work into your operation will be challenging. Changing how the organization thinks about risk is tougher. This is the primary reason for introducing change slowly, so that implementation and integration can go hand-in-hand. Although the implementation of the principles and practices of CRITICAL STEPS is straightforward, influencing the mindset that goes along with it takes time and dedication by the management team—follow-through. Before implementing CRITICAL STEPS principles and practices institutionally, it is advisable to integrate them into operations in four stages[9]:

1. *Exploration*—The central task here is to determine if there is a business need for change. The first step is to educate the executive leadership and senior managers on the principles and practices of managing CRITICAL STEPS. If interested, the organization should follow up with a needs-assessment, which is described later in this chapter. The information gleaned from the assessment, reviewed within the context of organization's performance history, defines the gap between the current state and a possible future state. Any change in the organization's philosophy, its organizational structures,

and ways of doing work requires validation with respect to its business mission—the organization's reason for existing. To help the management team make an informed decision, the *business case* (benefits, costs, risks, and assumptions) must be clear, and the methodology should consistently mesh with company values.[10]

Caution: Executive leadership must be thoroughly convinced of the value the management of CRITICAL STEPS would bring to the organization. Much of the task of integrating and implementing the principles and practices of managing CRITICAL STEPS into the organization depends on their dedication and follow-through.

2. *Preparation*—The intent of this stage is to develop a plan and establish leadership and responsibility for the initiative's integration and implementation. If committed to the initiative after the exploration stage, begin developing a change management plan—a playbook—for implementing CRITICAL STEPS into operations.[11] The playbook, described later, is updated and reviewed regularly through the pilot and full implementation phases. After identifying key stakeholders, conduct training. Recruit in-house champions, specialists, and local leaders. Remember to solicit input from frontline workers, the richest source of safety insights. It may be prudent to conduct these preparations in concert with the pilot, the third phase.

Caution: Incorporate the principles and practices of managing CRITICAL STEPS into your operation *slowly*, limiting the scope to assets vital to the organization's business mission. Limit your focus to one or two key assets initially, such as mission-critical products. We suggest not including personnel safety initially. After a year or so, when the organization becomes proficient in managing CRITICAL STEPS with one or two select assets, the organization can incorporate other assets associated with personnel safety, quality, and reliability.

3. *Pilot Implementation*—Conduct a small-scale, proof-of-concept pilot project in which a specific operational unit commits to incorporate CRITICAL STEPS into its operation. This stage has a dual purpose: 1) to decide if the initiative is worthwhile to implement for the enterprise (an experiment), and 2) to collect lessons learned from the operational unit's experience. Consider the Guidance for Implementation described later. Implementation should target the most significant and most frequently occurring operational risks. The practices of managing CRITICAL STEPS should be tailored to meet the needs of the unit. Achieve small wins early. This stage is characterized by frequent feedback and problem-solving at both the sharp and blunt ends of the organization.

Caution: Avoid selecting a troubled organizational unit for the pilot. You want to know if the CRITICAL STEPS principles and practices will help a healthy operational unit. Focus efforts where success is likely, and resistance is minimal.

4. *Full Implementation*—This phase involves scaling up the initiative to be integrated and implemented throughout the organization or enterprise. Revise the playbook to incorporate lessons learned from the pilot.

Institutionalize the principles and new practices of managing CRITICAL STEPS into the organization's management system. Organizational structures should be aligned to shape new practices consistent with the attributes and principles of H&OP.[12] Integrating and implementing CRITICAL STEPS should not be considered optional but as the new expectation for planning and executing work. Encourage systems thinking as the prevailing approach to learning and continuous improvement (realigning). Senior management regularly monitors the initiative's business case, its progress of implementation, and the integration of and adherence to the attributes and principles of managing H&OP.

Remember that implementation is about managing: 1) knowing the *gap* between where you are and where you want to be, 2) developing a *plan* to close the gap, 3) *implementing* the plan, 4) *monitoring* progress, and 5) *adjusting* as needed.[13] Integration is more about leadership, communication, and relationships.[14] Whether changing the way you perform work or the way you think about work, both require alignment of the organization to support its stakeholders.

GUIDANCE FOR IMPLEMENTATION

This section describes the set of management activities important to the implementation of the principles and practices of CRITICAL STEPS into operations. In addition to describing the generalized sequence of implementation activities, this section includes guidance on the planning and conduct of a needs-assessment of current practices associated with the management of CRITICAL STEPS in operations and the development and content of the playbook—the change management plan.

GENERAL SEQUENCE OF IMPLEMENTATION ACTIVITIES

The following activities identify and describe those actions for implementing the principles and practices of CRITICAL STEPS. Note that several could be pursued concurrently. Priority and emphasis of discrete implementation activities should be reviewed and then adjusted based on risk and business need.

Note: Because of the novelty of H&OP, RISK-BASED THINKING, CRITICAL STEPS, and systems thinking, it may be expedient for the organization to train and mentor internal H&OP specialists. These individuals would be able to guide and train the rest of the organization in its adoption of the principles and practices of CRITICAL STEPS.

1. *Gain sponsorship from senior management*—Develop and review the business case for implementing CRITICAL STEPS. Consider incorporating this initiative into the organization's strategic plan and assigning a resource sponsor and champion of the initiative—may be one and the same.

2. *Conduct a needs-assessment of current practices (to know where you are)*—Compare current practices and conditions against the benchmark criteria (described later in the *Needs-Assessment* subsection). This defines the performance gap between current and best practices.

3. *Develop a positive vision of CRITICAL STEPS principles and practices when they are fully integrated into operations*—Communicate the initiative and its objectives in a formal policy statement.

4. *Develop and implement a playbook (see Playbook subsection)*—Select the preferred level of implementation: pilot/full (see *A Four-Stage Approach* earlier). Coordinate the implementation of CRITICAL STEPS with other change initiatives. Resolve cross-purposes, vocabulary, and duplicate effort. Take advantage of strengths already in play, identified during the needs-assessment.

5. *Train the organization's line managers on the principles and practices of managing CRITICAL STEPS*—Using a systematic approach to training, conduct training that addresses the playbook (strengths and weaknesses), RISK-BASED THINKING, systems thinking, and the content from frontline workers (see item 14). Include a discussion of the initiative's business case, contrasting managers' roles in integration and implementation. This training should precede training for frontline workers by 2 to 3 months to give managers some soak time with the new principles and practices. Managers and supervisors must be able to reinforce and preserve existing positive practices as well as model and coach new practices and behaviors.

6. *Train line managers and first-line managers on the conduct of field observations and feedback*—Using a systematic approach to training, conduct field observation and feedback training. Training should provide a break-in period to allow practice. During the break-in period, align the organization to support time in the field and gathering and responding to feedback.

7. *Identify the most important, most frequent operational risks for the operational unit*—Referencing the facility's safety analysis report, the organization's event history, and latest risk register (if available), identify procedures associated with high-risk, high-frequency work activities, relevant to key assets. Explicitly denote CRITICAL STEPS and related RIAs. Develop a workdown curve to systematically track the review and revision of technical procedures. Include the workdown curve in the playbook.

8. *Identify perpetual CRITICAL STEPS*—Conduct CSM to pinpoint recurring CRITICAL STEPS, related RIAs, and needed defenses.

9. *Reconcile vocabulary*—H&OP and CRITICAL STEPS have a unique vocabulary. Clarify terms and phrases used in the organization to be consistent internally and universally understood. Reinforce the ongoing and correct use of terms and phrases. (The reader may adopt or revise the terms and phrases used herein or described in the Glossary (Appendix 1).)

10. *Pinpoint new behaviors*—These behaviors are necessary to identify and control CRITICAL STEPS and to fail safely. They are non-negotiable expectations (may include **Hu** Tools; see item 11). Develop these in collaboration with work groups expected to adopt the new behaviors. Conduct a

*local factor analysis** for each new behavior and work group; for example, prework discussions (new behavior) by supervisors and frontline workers (work groups) or field observations for managers.

Caution: Remember to identify recurring unwanted practices that would work at cross-purposes with new behaviors and to eliminate the conditions that enable them.

11. *Identify and develop **Hu** Tools*—If needed, develop working-level practices that promote positive control of CRITICAL STEPS and related RIAs. Solicit input from the work groups on the development of selected **Hu** Tools. Again, using a systematic approach to training, conduct hands-on training on select **Hu** Tools. In this training, incorporate multiple opportunities to practice the tools and to receive performance feedback. Integrate **Hu** Tools into technical training activities.

12. *Augment the adaptive capacity of frontline workers*—Building on what's already working well (usually workers' technical expertise), consider realigning other organizational functions in support of slack, RISK-BASED THINKING, flexibility, redundancy, teamwork skills, and situation awareness (see Chapter 6).

13. *Align the organization to integrate the principles and implement the practices of CRITICAL STEPS into operations*—Organizational functions affected include, as a minimum, engineering design, work management, procedure development, training, lock out/tag out (LO/TO), and CA/PA processes.

14. *Train frontline workers on CRITICAL STEPS*—Using a systematic approach to training, conduct training on principles and practices of CRITICAL STEPS. As a minimum, content should address the following elements:

    a. Risk: assets and their critical parameters, built-in hazards, and human fallibility

    b. Work execution process, emphasizing value addition and value extraction

    c. Loss of control and error traps

    d. Recognizing pathways and touchpoints

    e. CRITICAL STEP (emphasizing applicability to personal lives as well as to work)

    f. Contrast of human actions, touchpoints, and CRITICAL STEPS and their outcomes

    g. RIAs and relationships with CRITICAL STEPS, assets, and hazards

    h. Positive control of CRITICAL STEPS, including select **Hu** Tools for control of CRITICAL STEPS and RIAs

---

* Local factor analysis examines carefully the workplace conditions (local) that either enable or inhibit a particular behavior. These are created by various organizational functions. The local factor analysis method is described in Tony's previous book *Risk-Based Thinking* (pp. 168–172, 201–203).

    i. Adapting to unanticipated work situations and landmines; *work-as-done* versus *work-as-imagined*

    j. Failing safely, including RISK-BASED THINKING, chronic uneasiness, situation awareness, and conservative decision-making

    k. Rudimentary team and communication skills, including ongoing conversations about risk, what must go right and related preconditions for safety

    l. Learning: prework discussions, observations and feedback, and postwork reviews

Note: Give people time to learn and practice new behaviors without risk of embarrassment—a break-in period—to reduce their anxiety regarding their ability to learn and do them. Give workers opportunity to provide feedback on their usefulness in the workplace. Managers and supervisors must be able to speak the language of CRITICAL STEPS and to model and coach the new behaviors.

15. *Conduct field observations and feedback*—Direct line managers to observe work on a regular basis, commensurate with the tempo of high-risk operations. Use paired observations regularly to give managers and supervisors feedback on their effectiveness in conducting observations, enabling conversations, and providing feedback. In addition to receiving feedback, managers and supervisors must be able to reinforce, coach, correct, and console people. (See subsection, *Field Observations and Feedback*, in Chapter 6.)

16. *Develop a robust reporting system*—Regularly collect and resolve feedback from frontline workers about serious and repetitive differences between *work-as-done* and *work-as-imagined*. Workers must have no fear in describing reality. Make reporting fast, simple, and easy to use. Ask line managers to give an accounting of SYSTEMS LEARNING in response to the feedback.

17. *Implement CRITICAL STEPS for other organizational units*—Prepare to scale up the integration and implementation of CRITICAL STEPS at an enterprise level. Leverage observed successes (and shortcomings) to extend implementation to other work groups and disciplines, being careful to tailor playbooks to technologies, cultures, and risks involved.

NEEDS-ASSESSMENT

Before developing your playbook, first conduct a needs-assessment of your organization's current practices and operational conditions using the benchmark criteria listed in the following. With the aid of one or more trained H&OP specialists, review the organizational unit's current management and workplace practices and operational conditions regarding the management of CRITICAL STEPS. The assessment helps identify the needs of the organization—its strengths, weaknesses, opportunities, and threats (SWOT) associated with the unit's current operations. A SWOT analysis will help you organize the results of the assessment—the starting point of your playbook. The following assessment criteria can be used for comparison with current practices:

1. What constitutes a CRITICAL STEP for a particular work group or organizational unit is clearly defined and understood.
2. Unnecessary CRITICAL STEPS are minimized in engineering design and modification processes. If not automated, engineering processes adhere to human-centered design standards around high-risk human actions, minimizing the presence and influence of error traps. Engineering design adheres to inherently safer design and system safety principles. Known CRITICAL STEPS are explicitly identified in design documentation.
3. CRITICAL STEPS, related RIAs, and their controls are identified during the work planning and procedure development/revision processes and are explicitly denoted as such.
4. CSM or a similar analysis method is used to identify perpetual CRITICAL STEPS for high-risk operations that are typically guided by technical procedures.
5. Technically trained and qualified workers are assigned to perform high-hazard work.
6. Before starting work, frontline workers acknowledge the important assets involved in the assigned tasks and are mindful of their built-in hazards in the work to be performed. They know what they expect to accomplish and what to avoid for every high-risk operation.
7. Prework discussions are systematically conducted for high-risk work, including work activities with one or more CRITICAL STEPS.
   a. Known CRITICAL STEPS in the procedure are identified or are otherwise denoted.
   b. Ways of losing control are explored (review of error traps and operating experience).
   c. Means of positive control for each CRITICAL STEP, such as **Hu** Tools, and related RIAs are agreed upon, and individual responsibilities are assigned.
   d. Means of emergency communications are identified.
   e. STOP-work criteria are specified.
   f. Contingencies (means of failing safely) are explored for each CRITICAL STEP.
8. Adaptive capacities are reviewed for the work at hand enhanced by ongoing, interdisciplinary, technical conversations.
9. Workers avoid at-risk and unsafe practices and so-called "necessary violations" to get work done at CRITICAL STEPS and RIAs.
10. Regardless of the absence of any hazards, CRITICAL STEPS are *always* considered critical.
11. People exercise positive teamwork skills, make conservative decisions when assets are threatened, stop when unsure, and get help.
12. Line managers and supervisors regularly devote time to observe work first-hand and are able to model and coach expectations.

13. Postwork reviews are regularly conducted for high-risk work that involved CRITICAL STEPS.
    a. Significant differences between *work-as-done* and *work-as-planned* involving CRITICAL STEPS, dominant error traps, and landmines are identified and reported.
    b. Surprises, unfavorable workplace conditions, field adjustments, and related responses made around CRITICAL STEPS are reported.
14. Using a formal CA/PA process, line managers promptly respond to post-work review reports.
15. Line managers are accountable for the effectiveness and timeliness of corrective/preventive actions.

Additionally, you may want to expand the needs assessment to conduct a SWOT analysis for each of the seven principles of managing H&OP.

PLAYBOOK—A DYNAMIC GUIDE TO MANAGING CHANGE

In sports, before engaging in a contest, a coach prepares a playbook to optimize the team's offensive and defensive capabilities against its opponent, matching the team's talent with the opponent's talent. Similarly, the playbook guides implementation of the principles and practices of CRITICAL STEPS consistent with your organization's unique needs. An initiative to manage CRITICAL STEPS cannot be done ad hoc but must be guided and tracked systematically. As the business environment, technology, and values change, update the playbook to sustain its relevance. Finally, the playbook serves as an important tool for managerial accountability—implementation of CRITICAL STEPS principles and practices should not be considered optional. As a minimum, the playbook should contain the following entries:

1. *Vision of the future state*—when CRITICAL STEPS principles and practices have been fully implemented and integrated into operations; based on input from the needs-assessment
2. *Needs-assessment plan and results*
    a. Strengths and related opportunities
    b. Weaknesses and related threats to safety, quality, reliability, etc.
3. *Sequence of activities (described earlier)*—who does what by when; resource requirements budgeted*
4. *Assumptions*—explicit recognition of conditions taken for granted relevant to the initiative, and their possible effects
5. *New behaviors/practices/expectations to enable/start*
    a. Local factors analysis[15]—enabling and inhibiting
    b. Identification of needed organizational changes (alignment)
6. *Old behaviors/practices to inhibit/stop (unlearning)*
    a. Local factors analysis—enabling and inhibiting
    b. Identification of needed organizational changes (realignment)

---

* Budgeting to resource the initiative comes *after*, not before, developing the playbook.

7. *Means to augment adaptive capacity of frontline personnel to support in-field adjustments*

8. *Means to verify the systems are aligned to support principles and practices of CRITICAL STEPS in the workplace*[16]—integration:

   a. Work management systems (including LO/TO)

   b. Procedure and work instruction development and revision

   c. Engineering design processes

   d. RISK-BASED THINKING and chronic unease

   e. Training and technical expertise

   f. Field observation and feedback

   g. Reinforcement and reward plan (for new behaviors/expectations)

   h. Workplace conversations about risk

   i. SYSTEMS LEARNING (including postwork review, reporting, and the CA/PA process)

9. *Communication plan*—engagement of as many people as possible to strengthen a consistent message about the principles and practices (expectations) of managing CRITICAL STEPS using various forms and channels of communication

10. *Effectiveness reviews*—periodicity of manager review and accountability

CALL TO ACTION—THE BUSINESS CASE

We trust by now that you are informed, prepared, and perhaps eager to execute CRITICAL STEPS, implementing and integrating its principles and practices into your operation. The business case for managing CRITICAL STEPS is threefold. There is a strong *benefit* to not only safety, but also quality, reliability, productivity, and even profitability of your operation. There are *costs* to execute CRITICAL STEPS, involving the time and expense to adopt its principles and practices. Finally, but most importantly, there is a *risk* if your organization does not adopt the principles and practices espoused herein.

The principal benefit of managing CRITICAL STEPS to the organization is to multiply the successes of people's performance in the workplace by reliably creating value without losing control of the built-in hazards. Improving safety—protecting assets by rigorously managing CRITICAL STEPS—is a *profit multiplier*. It sustainably improves profits, by improving revenues through consistent quality and on-time delivery and by reducing the per unit expenses of production by avoiding the costs associated with events and other inefficiencies triggered by consequential errors.[17] And, by educating and training the frontline workforce on the principles and practices of managing CRITICAL STEPS, they will perform work more mindful of assets and their threats in the workplace.[18]

As with all substantial organizational changes, it will take time, training, structural changes, and related opportunity costs to implement the foregoing objectives. Adopting new ways of thinking—managing and doing work—is not a one-and-done affair. It takes practice and continual learning, individually and collectively. Depending on the size and complexity of the organization, executing CRITICAL STEPS may take as little as 6 months to a year for organizations with enlightened personnel, or 2–5 years for others until the practice of managing CRITICAL STEPS becomes the common way of doing work. Enlightenment occurs when the organization's line managers realize that performance and the management of human performance risk is a function of the system.

> **Caution**: Integrating and implementing CRITICAL STEPS into operations is not an initiative focused on fixing the worker. This is an initiative focused on managing the system.

The combinations of fallible human beings, imperfect designs, procedures based on faulty assumptions, difficult/risky tasks, complex technologies, the marshalling of hazards, and numerous regulatory requirements seem to make every aspect of operations seem critical. You must ask yourself and your management team a few introspective questions. What's the risk if the organization doesn't manage CRITICAL STEPS? How will CRITICAL STEPS focus the organization's safety efforts? What's the ultimate risk and subsequent cost to people and the business without a focused emphasis on making the right things go right? Does the benefit exceed the cost? Can you afford not to do it?

Think of it another way. Like many businesses, your time is money. Your success until now reveals your organization has a bias for action. Managing CRITICAL STEPS gives you greater clarity around those things vital to the organization's success, making sure they go right the first time. The process also discriminates between what is truly important to safety, quality, reliability, and productivity and what is nonthreatening. Knowing where "failure is not an option," you can leverage your bias for action to focus rigor where it counts the most—CRITICAL STEPS and their related RIAs. The workforce can be given greater agility to pursue efficiencies on non-critical activities, contingent on enabling and sustaining their technical expertise and chronic unease in the workplace and its organization. Your workforce can now operate more confidently, both at CRITICAL STEPS and where performance is not so critical.

> **Caution**: If not fully convinced, do not pursue an initiative to adopt the principles and practices of managing CRITICAL STEPS. This initiative requires a significant commitment and focus by the organization's executives and managers. Too often an initiative is implemented, not because the manager believes in its efficacy, but because it will impress others. This mindset leads to the "program of the month," and without commitment, follow-through, and accountability, it will fail.

If you believe that managing CRITICAL STEPS has the potential to improve the safety of your people, products, or property, there is much that can be done starting tomorrow. Chances for successful integration and implementation are optimized with the participation of your management team and the wisdom from the workforce's informal leaders. But keep it simple. Go slowly. There is no hurry. Adapt the guidance in this book to meet your unique needs. As your understanding and experience with this risk-based approach matures, you will take on a more integrated, systemic view of what must go right, applying CRITICAL STEPS to more assets. As Dr. Erik Hollnagel says in his book, *Synesis*, "it is not about productivity or quality or safety or reliability but about *all of these together*."[19]

KEY TAKEAWAYS

1. The executive leadership must be convinced of the value of CRITICAL STEPS before devoting resources to its implementation. This is best accomplished with a carefully prepared business case.
2. Integration is required to ensure an ongoing process of learning (and unlearning), of aligning the system to support both the systematic identification and control of CRITICAL STEPS.
3. When implementing the principles and practices of managing CRITICAL STEPS, the ongoing aims are to:
   a. Enable identification and control of CRITICAL STEPS.
   b. Enable adaptive capacity to respond to the unexpected.
   c. Fail safely, if practicable.
4. The execution of the principles and practices of CRITICAL STEPS, if it is to be sustainable, must become a way of thinking and doing work—of doing business. This is the purpose of integration.
5. The principles and practices of CRITICAL STEPS should be implemented slowly using the four-stage approach, limiting the scope to assets associated with the organization's business mission. The scope of the application of CRITICAL STEPS can be expanded as experience is gained with the risk-management process.
6. Continuous improvement and organizational alignment are optimized through systems thinking.
7. A needs assessment of current practices helps identify not only weaknesses to improve but also strengths to take advantage of in managing the human performance risks. This information is the starting point of the change management playbook.
8. An initiative to manage CRITICAL STEPS cannot be done ad hoc but must be guided and tracked systematically using a tool such as a playbook.

CHECKS FOR UNDERSTANDING

1. True or False. Integration focuses exclusively on developing compliance with procedures, expectations, and regulatory requirements.
2. Which organizational functions would be necessary consider in implementing CRITICAL STEPS? (Mark all that are correct.)
   a. Finance
   b. Maintenance
   c. Human resources
   d. Operations
   e. Management/administration
   f. Supply chain
   g. Engineering
   h. All the above
   i. None of the above

3. Fill in the blank. A business case considers the _____, _____, and _____ to the organization for the initiative in question.
4. A proof of concept should deploy CRITICAL STEPS in an organizational unit that is:
   a. Operations focused
   b. High-risk
   c. Generally successful
   d. All the above
5. True or False. A needs-assessment of current work practices relevant to CRITICAL STEPS should precede the development of a change management playbook.

THINGS YOU CAN DO TOMORROW

1. Explore the principles of managing H&OP and attributes that strengthen the integration of CRITICAL STEPS into operations. Discuss executives'/ senior managers' acceptance/resistance to their practical application to the business and its operations.
2. Using your organization's business planning process, develop a first draft business case for integrating the management of CRITICAL STEPS into operations. Review its assertions, facts, and assumptions. Solicit compliments, reservations, doubts, and suggestions.
3. Decide on one or two tangible assets, such as mission-critical products, as the initial focus of CRITICAL STEPS. We suggest not including personnel safety at the outset until the organization is proficient with identifying and controlling CRITICAL STEPS.
4. In collaboration with the management team, prepare a playbook to guide your execution of CRITICAL STEPS for the high-hazard operational units. Coordinate with other initiatives that may work at cross-purposes with CRITICAL STEPS. Invite respected, informal leaders from the workforce to participate.
5. Discuss with your colleagues or management team what success would look like in 2–3 years. What new behaviors would you see happening among frontline workers? Supervisors? Line managers? Executives? This will help you define the desired end state in your playbook.
6. List five strengths of your current organization that can be leveraged to promote and advance CRITICAL STEPS integration.

REFERENCES

1 Quote attributed to a talk given by Dr. Karl Weick during a 2013 High-Reliability Organizations conference hosted by Dow Chemical in Midland, Michigan, USA.
2 Bossidy, L., and Charan, R. (2002). *Execution: The Discipline of Getting Things Done.* New York: Crown (p. 89).
3 Ibid. (p. 15).
4 Hollnagel, E. (2021). *Synesis: The Unification of Productivity, Quality, Safety, and Reliability.* Abington: Routledge (pp. 1–4).

5 Ibid. (p. 93).
6 Hollnagel, E. (2014). *Safety-I and Safety-II: The Past and Future of Safety Management.* Farnham: Ashgate (p. 127).
7 Herrera, I., Lay, E., and Cardiff, K. (2017, June). 'From Air to Ground—Resilience Strategies and Innovation Across Critical Infrastructures.' *Proceedings of the 7th Resilience Engineering Association Symposium.* Retreived from: https://www.resilience-engineering-association.org/resources/symposium-proceedings/.
8 Covey, S. (2015). *The 7 Habits of Highly Effective People.* New York: Mango Media (p. 47).
9 Turner, K., Renfro, C., Ferreri, S., and Shea, C. (2018). *A Staged Approach to Implementing Change.* Chapel Hill, NC: UNC Center for Medication Optimization. Retrieved from: https://changemgmt.media.unc.edu/stages/.
10 Muschara, T. (2018). *Risk-Based Thinking: Managing the Uncertainty of Human Error in Operations.* New York: Routledge (pp. 217–218).
11 Ibid. (pp. 218–223).
12 Hopkins, A. (2019). *Organizing for Safety: How Structure Creates Culture.* Sidney: Wolters Kluwer (pp. 28, 39).
13 Muschara, T. (2018). *Risk-Based Thinking: Managing the Uncertainty of Human Error in Operations.* New York: Routledge (p. 223).
14 Schein, E. (2004). *Organizational Culture and Leadership* (3rd ed.). San Francisco: Jossey-Bass (pp. 246–271).
15 For a detailed description of *local factor analysis,* refer to Muschara, T. (2018). *Risk-Based Thinking: Managing the Uncertainty of Human Error in Operations.* New York: Routledge (pp. 201–203).
16 For a detailed explanation of organizational alignment, see Muschara, T. (2018). *Risk-Based Thinking: Managing the Uncertainty of Human Error in Operations.* New York: Routledge (pp. 125–132).
17 Van Dyck, C., Frese, M., Baer, M., and Sonnentag, S. (2005). 'Organizational Error Management Culture and Its Impact on Performance: A Two-Study Replication.' *Journal of Applied Psychology,* 90(6): 1228–1240. https://doi.org/10.1037/0021-9010.90.6.
18 IDC Research (2008). *$37 Billion: Counting the Cost of Employee Misunderstanding.* (whitepaper). Framingham, MA: IDC Research.
19 Hollnagel, E. (2021). *Synesis: The Unification of Productivity, Quality, Safety, and Reliability.* New York: Routledge (p. i).

Epilogue: People Have Dignity and Value

We hold these truths to be self-evident, that all men are *created* equal.[1]

H&OP's number one principle is that **"People have dignity and inherent value as human beings."** We are equally valuable, yet possess diverse talents and roles. Human error is not sin. Yet, too often, society responds to the mistakes of others as if they were immoral choices, and punishment ensues proportionate with the severity of the event. The dignity of persons must be *preserved*; responses to mistakes and violations are important. Accountability can and should be exercised without robbing people of their dignity. Violating this principle poses a serious obstacle in achieving open and effective communications.[2] To avoid violating this principle, you must accept the gravity of the following assertions:

> Every human being is *intrinsically* valuable.
> Every human being is *equally* valuable.
> Every human being is *exceptionally* valuable.

Human beings are not valuable because of their usefulness to serve the desires of others. We all have innate value simply because we are human; we are not inanimate objects. No one person's life has more value than another. Everyone has exceptional value in that we each have a mind and a conscience, able to reason, knowing right and wrong, all essential for robust conversations about risk in the workplace.[3]

Human error breaks things; sin, with its selfish tendencies, breaks relationships. Relationships are essential to open communication, and, therefore, everyone, regardless of position, must be respected as human beings with ordinary human needs, values, and limitations. Injustices in the workplace are usually not due to inferior performance as much as they are the outcome of bad relationships.[4]

There are no laws against treating people with dignity, respect, fairness, and honesty, characteristics that are important to building trust and communication within any organization. People, especially frontline workers at the sharp end, should be treated not as a liability—as objects to be controlled—but as knowledgeable and respected agents of the technical side of the organization who have its best interests at heart.

It is our sincerest belief that human dignity arises not from ethnicity, creed, color, gender, national origin, education, or a person's business value, but from our status as image bearers, as a creation and reflection of the triune God.

> it is *with the awe and the circumspection* proper to them, that we should conduct all our dealings with one another, all friendships, all loves, all play, all politics. **There are**

DOI: 10.1201/9781003220213-9

no ordinary people. You have never talked to a mere mortal. . . . Next to Jesus Christ, your neighbor is the *holiest* object presented to your senses.[5]

—**C.S. Lewis**
The Weight of Glory

NOTES

1 The first line of the *United States Declaration of Independence.*
2 Muschara, T. (2018). *Risk-Based Thinking: Managing the Uncertainty of Human Error in Operations.* New York: Routledge (p. 210).
3 Ensor, J. (2012). *Answering the Call: Saving Innocent Lives, One Woman at a Time.* Peabody, MA: Hendrickson (p. 39).
4 Dekker, S. (2007). *Just Culture: Balancing Safety and Accountability.* Aldershot, UK: Ashgate (p. 142).
5 Lewis, C. S. (1949). *The Weight of Glory: And Other Addresses.* New York: HarperCollins (p. 45).

Appendix 1: Glossary of Terms

If you want to change the way you think, change the words you use.

—Karl Weick

Human and organizational performance (H&OP) is an emerging technology in the application of safety science. There is a vocabulary associated with it and the management of risks associated with human performance during operations. **You cannot manage what you do not understand.** This vocabulary improves your understanding of H&OP, allows communication about related risks, and aids your ability to manage them. Language is fateful—it influences outcomes. That means definitions are important, which is why this glossary is provided. It will serve as a ready reference as you read this book.

Accident	An unplanned, surprise occurrence that results in injury, damage, or loss, usually occurring within moments of a trigger.
Accomplishment	A product of work behavior that delivers value toward achieving the organization's mission; work output that remains after behavior.
Action	A physical, bodily motion (application of force), e.g., handling, movement, modification, or alteration of an object; or an expression of sound.[1]
Active Error	An error (action) that unintentionally triggers a loss of control of a hazard, altering the state of an asset and resulting in its immediate harm.
Adaptive Capacity	The ability to change when circumstances change; readiness to adjust to challenges to make things go right.[2]
Anticipate	To *know what to expect for the work-at-hand*—to foresee the behavior of the system and implications of a proposed act or course of actions. Includes looking ahead for possible pitfalls and potential consequences to assets, given the hazards and their pathways, exemplified by asking "what if . . .?"
Asset	Something of high value to an organization that could sustain injury, damage, or loss and is important to an organization's mission, survival, and sustainability. Examples:

- People (health and well-being—the most important asset)
- Product (quality)
- Property (facilities, tools, and equipment)
- Environment (earth, water, and air)
- Time (productivity and schedule)
- Money (cash, capital)
- Proprietary information (trade secrets)
- Software
- Intellectual property
- Public trust and reputation

Barrier	Means used to protect an asset from harm by limiting or impeding uncontrolled transfers of energy (e.g., electrical, chemical, heat, kinetic), movements of matter (e.g., loads, shipments, product, fluids), or transmissions of information (e.g., passwords, signals, digital records, software).

Behavior
The mental and physical efforts to perform a function; observable (moving or speaking) and hidden (perceiving, thinking, analyzing, deciding) tasks or activities by an individual.

Blunt End
The people of the organization who are removed in time and space from direct production activities, who organize and control its activities, set policies and priorities, manage business processes, decide on technical design, provide resources, and prepare procedures. People who influence (directly and indirectly) the production work of frontline workers.

Business Case
The reasoning, usually documented, to assess the benefits, costs, risks, and options for a proposed new and significant activity. Reasoning for a proposed course of action that is aligned with the organization's purposes and values.

Chronic Unease[3]
Generally, the experience of concern about risks, exemplified by a healthy skepticism about one's decisions and the risks inherent in work environments. Operationally, a deep-rooted respect for the technology, its complexities, and its built-in hazards that spawns an ongoing:

- Mindfulness of the presence of hidden threats in the workplace, such as landmines and error traps; vigilance
- Readiness for surprises, understanding that work can be characterized as VUCA—volatile, uncertain, complex, and ambiguous; pessimism.
- Preoccupation with impending or current transfers of energy, movements of matter, or transmissions of information that could threaten the safety of an asset; readiness
- Sensitivity to human fallibility—the capacity to err (requires humility)
- Reluctance to consider any job or workplace situation as routine; skepticism

Complex Adaptive System
A performance system characterized by the presence of many components (some individually adaptive, such as people), possessing concurrent and possibly obscure interdependencies—not easily comprehensible to any one person—that sometimes cause the system to operate in unanticipated ways, producing surprising outcomes.

Conservative Decision-Making
A decision that places greater value on an asset's safety than on achieving the immediate production goal, resulting in actions to place the asset in a known safe condition. The intent is to preserve or improve an asset's margin of safety.

Control
Means that guide, coordinate, or regulate behavior choices, promoting desired action and outcome while improving the chances of error-free performance. Sometimes used as a synonym for a defense, as in an engineered "control."

Critical Parameters
An asset's set of key variables that define its integrity (safety); if exceeded, irreparable harm is suffered. The collection of critical parameters defines the limits of the asset's safe operating envelope.

CRITICAL STEP
A human action that will trigger immediate, irreversible, and intolerable harm to an asset if that action or a preceding action is performed improperly.

Danger
An asset's exposure or proximity to one or more hazards that can cause harm to the asset; created when a pathway is created.

Defense
Means taken to protect an asset against harm. Methods used to control or reduce an asset's exposure to a hazard, such as controls, barriers, and safeguards. Anything that tends to reduce the frequency or severity of harm.

Defense-in-Depth	The overlapping capacity of redundant defenses to protect an asset from danger, such that a failure of one defense is compensated for by other defenses, thereby avoiding harm.
Energy	The ability (capacity) to do work, such as heat, mechanical, gravitational (potential), kinetic, chemical, electrical, and nuclear. It is the uncontrolled release of energy that is typically dangerous to humans and other assets.
Error	A behavior that unintentionally deviates from a preferred behavior for a given situation. An act, an assertion, or a belief that unintentionally deviates from what is correct, right, or true.[4] Sometimes error involves a *loss of control*, but usually not recognized as an error by those involved until after an unwanted consequence is realized.
Error Trap	An unfavorable condition in the workplace or in the mind of the performer— in the here and now—that either creates uncertainty or otherwise enhances the chances of losing control during the performance of a task.
Event	An unwanted occurrence involving harm (injury, damage, or loss) to one or more assets associated with an uncontrolled transfer of energy, movement of matter, or transmission of information.
Expert Intuition	Unconscious recall from memory; awareness triggered by a situational cue of accumulated experience and an extensive knowledge base.
Fail Safely	The capacity to limit harm done—assuming there is a loss of control of an operational hazard—to place an asset in a safe state after a failure or fault, before the onset of significant harm, if possible. This includes the capacity to recover from either a loss of control or harm realized.
Failure	Ceasing to function or deviating from specification; losing control of a hazard or one's action; harm to an asset; not accomplishing an expected result, in some cases despite doing what was considered appropriate.
Fast Thinking	An effortless approach to thought, quickly making intuitive decisions with little or no deliberate attention, often automatically with no obvious awareness of voluntary control; subject to experiences, beliefs, emotions, and instinct.
Harm	Injury or damage to an asset; or loss of value, capability, or impairment of a mission—tangible or intangible. An unwanted change in the key characteristics (critical parameters) or state of an asset.
Hazard	A source of energy, matter, or information that could harm an asset or cause its loss—usually built into a facility for operational and business purposes. Any condition in the workplace that could harm or trigger harm to an asset.
High-Reliability Organization (HRO)	An organization that operates successfully and persistently in an unforgiving physical, social, or political environment, rich with the potential for error, and where complex processes are used to manage a complex technology to ensure success and avoid failure.[5] Sometimes HRO is used to denote the processes of *high-reliability organizing*.
Human Error	An error that refers to the slips, lapses, fumbles, and mistakes of humankind, regardless of whether one's goal is accomplished or not—not all human errors have bad outcomes.

Human and Organizational Performance (H&OP)	The practical application of safety science; the collective performance of an organizational unit involving the behavior choices and work outputs of several workers performing within the systemic context of the organization's social and technical environments.
Human Performance (Hu)	The behavior (B) of an individual to accomplish a specific result (R); (Hu → B + R).
Human Performance Tools (Hu Tools)	Discrete mental and social skills that complement an individual's technical skills to achieve safe and efficient task performance, carving out time for thinking before doing. May be referred to as error prevention tools or non-technical skills.
Improper	Descriptive of human actions, whether in error or otherwise irregular, that may involve a loss of control of hazards or the creation of unsafe conditions.
Job	A combination of tasks and duties (responsibilities) that define a particular position within an organization.
Landmine	A metaphor for an undetected hazardous condition in the workplace that is poised to trigger harm. A workplace condition that increases the potential for an uncontrolled transfer of energy, matter, or information with one action—an unexpected source of harm—unbeknownst to the performer. Includes compromised or missing defenses; also known as an "accident waiting to happen."
Latent Condition	An unknown situation or condition established earlier that persists, whether in the management system, production system, or the workplace, which may lie dormant for long periods of time, doing no apparent harm.
Latent Error	An error (action, inaction, or decision) that creates a potentially unsafe (latent) condition, unnoticed at the time, causing no immediate, apparent harm to an asset, but that could combine with other errors, occurrences, or conditions later to realize the harm.
Learn	A change in behavior in response to the acquisition of new knowledge and an understanding of its meaning and application. In the context of Risk-Based Thinking, *knowing* requires active learning to understand:

1. *What has happened* (previous experiences relative to the task at hand)
2. *What is happening* (situation awareness of the task at hand)
3. *What to change* (identification of behavior and system changes needed to enhance the safety of assets and the resilience of the organization for future similar operations)

Learning Team	An ad hoc method of organizational learning that creates a deeper, context-rich understanding of how work really happens by taking advantage of those who know and perform the work in question.[6]
Line of Fire	Physical path that energy or an object would travel if stored energy were released uncontrolled; an asset's exposure to harm while in the path.
Local Factors	A set of workplace conditions in the here and now that influence the behavior choices of an individual performing a task or taking an action.[7] Sometimes referred to as "performance-shaping factors" or "error precursors."
Loss	The reduction in the value of an asset (usually monetarily), or the physical deprivation of an asset.

Loss of Control	Lack of necessary regulation of the flow, intensity, and magnitude of energy, matter, or information, resulting in an unintended departure from expectation.
Management System	A formally established and documented set of activities designed to produce specific results in a predictable, repeatable manner; commonly referred to as a "program."
Mental Model	A person's or group's mental organization of knowledge about how something works; a mental picture or other representation of the arrangement and interaction of elements of a system that describes the basic ways in which something functions.
Mentoring	A long-term relationship between two people in which a more experienced or more knowledgeable person passes wisdom on to a less experienced or less knowledgeable person.[8,9]
Monitor	Attention devoted to a particular thought or object—especially the safety of assets and their related built-in hazards—comparing actual status with the desired status. In the context of RISK-BASED THINKING, concentration on varying levels of risk of harm to assets, such as: • The emergence of pathways for transfers of energy, movements of matter, or transmissions of information, especially for important assets • CRITICAL STEPS and related Risk-Important Actions for the work at hand • Important changes in an asset's critical parameters to allow timely response to avoid harm if needed • Trends in key indicators of safety and resilience (leading and lagging)
Organization	A system, comprised of people, resources, and technology, designed to coordinate and direct a group's individual and collective behaviors toward the accomplishment of its mission.
Organizational Factors	Conditions, processes, and practices in the upper echelons of an organization that create and moderate workplace conditions that, in turn, influence people's behavior choices during work. Examples are: • Management and leadership practices • Management systems • Procedure development and revision • Training programs • Operational work processes • Espoused values and priorities • Structures, systems, and components of technology • Type and availability of resources
Pathway	An operational situation in which an asset's transformation (change in state) is poised to occur by either a transfer of energy, a movement of matter, or a transmission of information. Exposure of an asset to the potential for harm for which only one action (equipment or human) is needed to alter the asset's state, for good or for bad.
Positive Control	A person's attempt to ensure that what is intended to happen is what happens and that is all that happens.
Principle	A fundamental truth about reality that cannot be reasoned from any other proposition—all other conclusions and practices in a particular domain are based on the one central idea.
Process	A series of activities or functions organized into tasks to produce a product or service. Means established to direct the behavior of individuals in a predictable, repeatable fashion as they perform various tasks.

Resilience   The intrinsic ability of a system to adjust its functioning before, during, and after a challenge, disturbance, or failure to sustain operations under both expected and unexpected conditions.[10] The ability of a system or process to return to a desired and acceptable (perhaps even better) state following an upset with little or no harm to its assets. The ability to succeed under varying circumstances, including the capacity of frontline workers to adapt to changing risk conditions in the workplace.

Respond   Knowing what to *do* 1) to protect assets from harm, 2) to recover from a harmful situation, or 3) to create a better outcome. In the context of Risk-Based Thinking, *respond* involves enhancing the frontline workers' capacity to adapt to changing risk conditions—especially during surprise situations—to protect assets from harm, such as:

* Exercising positive control of CRITICAL STEPS (whether planned or unplanned)
* Building slack or buffers into operations or tasks through resources, time, alternatives, flexibility, expertise, contingencies, etc.
* Taking actions to 1) eliminate the task or operation, 2) prevent error, 3) catch an error before it triggers harm, 4) detect a defect, or 5) mitigate harm done

Risk   The potential harm (injury, damage, or loss) an asset could experience when exposed to a hazard (pathway) while under the influence of a human action (touchpoint).

RISK-BASED THINKING   Four interdependent habits of thought (anticipate, monitor, respond, and learn) in the conduct of work that enhance a person's readiness and the capacity to adapt by building work-specific knowledge about the safety of assets involved in the work activity.

Risk-Important Action (RIA)   A reversible human action preceding a CRITICAL STEP that either:

* Create or remove pathways for the transfer of energy, movement of matter, or transmission of information that expose an asset to a hazard (necessary for the conduct of work)
* Increase or decrease the number of actions needed to begin work
* Influence the presence or effectiveness of barriers or safeguards that protect assets, or
* Influence the ability to maintain positive control of the moderation of hazards at CRITICAL STEPS

Risk-Important Condition   A precondition established by a Risk-Important Action that is necessary for:

* Work (a pathway for energy, matter, or information), or
* Positive control of CRITICAL STEPS (control of a touchpoint), or
* Failing safely if a loss of control occurs at a CRITICAL STEP (barriers and safeguards)

Safeguard   Means of mitigating or minimizing the harm done to an asset after the onset of injury, damage, or loss.

Safe Operating Envelope (SOE)   The boundaries and related safety margins for safe operation as defined by an asset's critical parameters.

Safety   An asset's freedom from unacceptable risk of harm, while ensuring what must go right indeed goes right, an emergent property of worker actions, the system and its internal dependencies.

Severity   The degree of harm realized (or possible) governed by the intensity or magnitude of the energy transferred, the amount of matter moved, the sensitivity of the information communicated, and asset's fragility at the time of the event.

Sharp End   Individuals in an organization in direct contact with production assets, safety-critical processes, and built-in hazards, who perform hands-on work to achieve the organization's purposes.

Situation Awareness The appreciation and accuracy of a person's current knowledge and understanding of working conditions compared to actual conditions at a given time.

Skill-of-the-Craft A set of knowledge and skills related to certain aspects of a task or job that an individual is expected to know and be fluent in without needing written or verbal guidance.

Slow Thinking   A conscious, analytical approach to decision-making; mindfulness characterized by attention to details and orderly concentration on an issue; subject to knowledge, prudence, and rationality.

System   A collection of components arranged in a particular way that function together to produce predictable, repeatable outcomes (e.g., an automobile).

Systemic   A characteristic of a system's structure or operation that can influence the choices of all parts or individuals in that system at the same time—in the here and now—instead of individual members or parts (not to be confused with "systematic," which means methodical).

Systems Thinking   An understanding of how systems behave and how their components interact with other components (and with other systems) to influence 1) people's choices and their outcomes in the workplace and 2) the resilience (defenses) of hazardous operational processes.

Task   A planned work activity with distinct start and stop points made up of a series of intervening actions of one or more people to accomplish one or more objectives, usually directed by a procedure or skill-of-the-craft. Sometimes used to denote a single, distinct action.

Touchpoint   A human interaction with an object (assets or hazards) that changes the state of that object through work. Work usually initiates a transfer of energy, a movement of matter (solid, liquid, or gas), or a transmission of information.

Uncertainty   Being unsure of what to do—a feeling of doubt. Ambiguity about the current or future states of an object or situation, usually based on insufficient knowledge or the complexity of the situation.

Verification   To confirm an object's (or component's) current state or condition.

Work   The application of physical strength or mental effort to achieve a desired result, whether a force over a distance or detailed reasoning, usually for the purpose of adding value.

Workaround   A deficiency in a procedure, component, or workplace that compels a frontline worker to take an adaptive, compensatory action—usually manual—different from the guiding procedure, to achieve one or more goals.

Work-as-Done   Work as it is accomplished by workers in the workplace.

Work-as-Imagined   Work as envisioned, planned, or believed to be happening by managers, designers, and procedure writers. Sometimes used alternatively as *work-as-planned*.

Work Execution   The process of accomplishing production activities that involve 1) preparation for work, 2) execution of work, and 3) learning from the work accomplished.

Worker (Frontline)	An individual in the organization who performs physical, hands-on work, having direct contact with (capable of altering the condition of) equipment, assets, built-in hazards, and safety-critical processes. Also, referred to as a frontline worker.
Worker (Knowledge)	An individual who works primarily with information, or one who creates and communicates knowledge for use by others.
Workplace	The physical location (local) where people can touch and alter assets and equipment or interact with built-in hazards and safety-critical processes.

NOTES

1 Sanders, M., and McCormick, E. (1993). *Human Factors in Engineering and Design* (7th ed.). New York: McGraw-Hill (p. 18).

2 Dekker, S. (2019). *Foundations of Safety Science: A Century of Understanding Accidents and Disasters*. Boca Raton: CRC Press (pp. 391–392).

3 Fruhen, L. (2015). 'Chronic Unease a State of Mind for Managing Safety.' [White paper]. Perth: Centre for Safety, University of Western Australia. The term was coined by Dr. James Reason (1997). *Managing the Risks of Organizational Accidents* (p. 214).

4 American Heritage Dictionary. Retrieved from: www.yourdictionary.com/error# americanheritage.

5 Rochlin, G. I. (1993). 'Defining "High Reliability" Organizations in Practice: A Taxonomic Prologue.' In: Roberts, K. (ed.). *New Challenges to Understanding Organizations*. New York: Macmillan (pp. 11–32).

6 Edward, R., and Baker, A. (2020). *Bob's Guide to Operational Learning*. Sante Fe, NM: Pre-Accident Investigation Media (pp. 137–138).

7 Reason, J. (1998). *Managing the Risks of Organizational Accidents*. Aldershot, UK: Ashgate. (p. 11).

8 Santora, K., Mason, E., and Sheahan, T. (2016). 'A Model for Progressive Mentoring in Science and Engineering Education and Research.' *Innovative Higher Education*, 38: 427–440. https://doi.org/10.1007/s10755-013-9255-2.

9 Maxwell, J. (2008). *Mentoring 101: What Every Leader Needs to Know*. Nashville, TN: Thomas Nelson.

10 Hollnagel, E. (2012, May 7). *How Do We Recognize Resilience?* Presentation at University of BC School of Population and Public Health Learning Lab.

Appendix 2: Managing Human and Organizational Performance—A Primer

This appendix provides a background on Human and Organizational Performance (H&OP) as described in Tony's first book, *Risk-Based Thinking: Managing the Uncertainty of Human Error in Operations*. This primer lays the foundation for managing the risk human performance poses to assets in an industrial, operational environment, namely: 1) the cornerstones of RISK-BASED THINKING (habits of thought), 2) the building blocks of managing H&OP, and 3) the principles (core beliefs) of managing H&OP. It is essential to have an appreciation of these principles and practices as one reads this book on managing CRITICAL STEPS.

CORNERSTONES OF RISK-BASED THINKING

RISK-BASED THINKING is referenced frequently in this book. Although not described in detail within these pages, it is important for the reader to appreciate the basic concept of RISK-BASED THINKING. RISK-BASED THINKING applies to all human functions at any organizational level: plant operations, management practices, construction, maintenance, engineering design, all the way up the corporate ladder to the board room. Researchers and academics in resilience engineering identified four cornerstone organizational practices important for success and safety. The exercise of control and the minimization of harm are undertaken by creating risk-relevant operational knowledge through the following *habits of thought*, which we collectively refer to as RISK-BASED THINKING.

1. *Anticipate*—know what to expect.
2. *Monitor*—know what to pay attention to.
3. *Respond*—know what to do.
4. *Learn*—know what has happened, what is happening, and what to change.

RISK-BASED THINKING encourages people to reflect on—to know—their work proactively, deliberately, and logically—not in a rush—to make things go right instead of mindlessly letting things happen to them. Frontline personnel in operations, maintenance, and administrative capacities perform CRITICAL STEPS and RIAs every day, and most of the time things go right. As a fundamental first principle of operational human performance, anyone can apply these habits of thought, anywhere, for any activity, any time. When applying RISK-BASED THINKING, people adjust their performance in response to disturbances, changes, and even opportunities to not only reduce the number things that go wrong (error and harm) but also increase the number of things that go right (at CRITICAL STEPS). The more likely that something will go right, the less likely that it will go wrong.

THE BUILDING BLOCKS OF MANAGING H&OP

We believe that people and the organizations they work in must be managed collectively as a performance system. This approach to managing operations, safety, quality, reliability, and production is commonly known as *human and organizational performance* (H&OP). H&OP is the practical application of safety science. It's a risk-based management philosophy, which recognizes that error is part of the human condition and that organizational processes and systems greatly influence employee behavior choices and actions and their likelihood of success.

Traditionally, managers have attempted to manage human performance simply by charging the members of the workforce to avoid mistakes—to pay attention and follow procedures. In contrast to this limited strategy, we offer an alternative approach to managing workplace human performance: *manage the risk* human performance (human fallibility) poses to your assets during operations. Strategically, the *core functions* of this approach, the main points of leverage for managing human performance risk, are described in the following. As illustrated in Figure A.1, managing H&OP involves three core operational functions (denoted by SMALL CAPS) and three

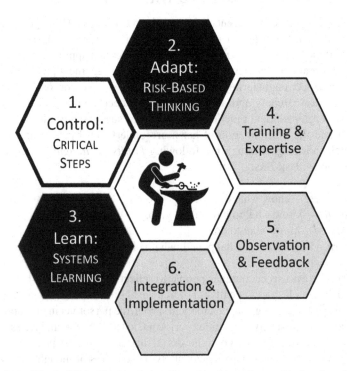

FIGURE A.1 The building blocks of managing human and organizational performance. This model depicts the six most influential organizational functions for managing the risks of human performance in the workplace.

management support functions. The core operational functions that involve daily risk management are:

1. *CRITICAL STEPS*—ensuring the right things go right when and where high-risk work is performed; the positive control of *high-risk human actions* involving impending transfers of energy, movements of mass, or transmissions of information between hazards and assets

2. *RISK-BASED THINKING*—adapting to changing and emerging risks in the workplace; as an *operating philosophy*, RISK-BASED THINKING augments the capacity of frontline workers to better recognize changing operational risks and to adjust their behaviors appropriately. It is not intended to be a program, but a way of thinking. Additionally, people possess a deep-rooted respect for the technology and its intrinsic hazards as well as an on-going mindfulness of impending transfers of energy, movements of mass, and transmissions of information—chronic uneasiness.

3. *SYSTEMS LEARNING*—the detection of systemic vulnerabilities and faulty defenses and the realignment of related organizational systems. SYSTEMS LEARNING involves the ongoing identification and correction of system-level weaknesses in an organization that 1) adversely influence behavior choices in the workplace and 2) tend to diminish the effectiveness of built-in defenses or inhibit the organization's resilience to challenges/threats (known/unknown). The effectiveness of SYSTEMS LEARNING depends on the management team's ability to apply *systems thinking*—understanding of how behavior and outcomes emerge from the system people work in.

Cells 4, 5, and 6 are important management practices that support the effectiveness of core functions 1 through 3. These are:

4. *Training and Expertise*—ensuring frontline personnel possess technical knowledge and skill to exercise RISK-BASED THINKING and to recognize and control CRITICAL STEPS; the systematic development of the workforce's understanding of the operation's technology and the nurture of a deep-rooted respect for its intrinsic hazards, essential for application of RISK-BASED THINKING and control of CRITICAL STEPS.

5. *Observation and Feedback*—creating operational opportunities for individual learning for frontline workers and SYSTEMS LEARNING for line managers; the real-time monitoring of shop-floor operations by line managers and supervisors to see firsthand what frontline workers actually do—the real-time choices they are making in the workplace.

6. *Integration and Implementation*—enabling: 1) RISK-BASED THINKING as a way of thinking, doing work, and doing business; and 2) managerial accountability for the execution of H&OP; the systematic and disciplined management of risks associated with H&OP (a management function), while enabling RISK-BASED THINKING and a chronic uneasiness in all facets of operations—sometimes referred to as "operational discipline."

PRINCIPLES OF MANAGING H&OP—CORE BELIEFS

Human performance risk and the complex systems people work in cannot be managed simply by design, procedure, supervision, or decree. The foregoing takes care of the known risk, but *built-in adaptive capacity* is needed to address unknown risks. Hence, managing H&OP is, by necessity, principle based. We believe managers must adhere to a set of core principles, fundamental truths about humans and organizations that are essential to the successful management of human performance risks. Notice that principles are not rules, procedures, or even practices. The organization's executive leadership team as well as its line management must believe in their hearts the validity and veracity of these principles. This book builds on the following principles.

1. *People have dignity and inherent value as human beings*—Everyone wants to be treated with respect, fairness, and honesty, characteristics that are important to building trust and communication within any organization. People should be treated not as a liability—as objects to be controlled—but as knowledgeable and respected agents of the technical side of the organization who have its best interests at heart. Relationships are integral to open communication; therefore, workers and employees must be respected as human beings with ordinary human needs, values, and limitations.

2. *People are fallible*—To err is human; error is normal. Fallibility is a permanent, inherent feature of the human condition. Human fallibility can be moderated, but it cannot be eliminated. It introduces uncertainty into any human endeavor, especially hands-on work. However, people are also flexible. They possess a wide range of capabilities and can adapt to accommodate inadequate resources, weak training, poor tools, schedule conflicts, and process shortfalls, among other workplace vulnerabilities. To protect assets, they can and often do adjust what they do. Assume people will err at the most inopportune times. Then, manage the risk.

3. *People do not purposefully come to work to fail*—Most people want to do a good job—to be winners, not losers; to apply and use their knowledge and skills responsibly, performing well. Error is not a choice—it's unintentional. Nobody errs purposely. Error is not sin—it's not immoral. Error tends to break things. Sin, in contrast, is selfish in nature and tends to break relationships. Reprimanding people for error serves no benefit. People are goal-oriented, and they want to be effective. They adapt to situations to achieve their goals. This means well-meaning people will take shortcuts, now and then, if the perceived benefit outweighs the perceived cost and risk. This, too, is normal. All people's actions, good and bad, are positively reinforced, usually by their immediate supervisors and by personal experiences of success (as they perceive it), which sustains their beliefs about what works and what does not.

4. *Behavior choices are predictable and manageable*—Human error is not random. It is systematically connected to the work environment—the

nature of the task and its immediate environment—local factors governing the individual's performance. Despite the certainty of human error over the long period for large populations, a specific error for an individual performing a particular task at a precise time and place under certain local conditions (here and now) can be anticipated and avoided. For example, what is the likely error when writing a personal check on January 2 of every year? You know the answer. However, not *all* errors can be anticipated or prevented. This is why defenses are necessary.

5. *Risk is an inherent, dynamic property of the way an organization operates*—When work is executed, various built-in sources of energy, tools, and material are used to accomplish the work. Consequently, hazards exist (built in) within an organization's facilities because of its purposes. For larger, complex organizations, risk is dynamic and lurks everywhere, and it varies as an outcome of the diverse ways an organization is designed, constructed, operated, maintained, and managed, as well as its tempo of operations. Safe and resilient organizations are designed and built on the assumptions that: people will err, things are not always as they seem, equipment will wear out or fail, and not all scenarios of failure may be known before operations begin.

6. *Organizations are perfectly tuned to get the results they are getting*—People can never outperform the system that bounds and constrains them. All organizations are aligned internally to influence the choices people make and the outcomes they experience—good and bad. All work is done within the context of its management systems, its technologies, and its societal, corporate, and work-group cultures—organizational factors. Organizations comprise multiple, complex interrelationships among people, machines, processes, systems, and various management systems; and managers go to great lengths to create structures for controlling work. But as we all know, there is no such thing as a perfect human, perfect system, perfect process, or a perfect procedure. Once systems go into operation, they prove to be imperfect.

7. *Conditions that spawn tomorrow's events exist today*—The conditions necessary for harm to occur always exist before the harm is realized. Some are transient; most are longstanding. Most of the time these conditions are hidden or latent. Latent conditions tend to accumulate within an organization and pose an ongoing threat to the safety of assets. Events occur when a loss of control or human error combines with one or more of these system weaknesses. These conditions are shaped by weaknesses at the organizational and managerial level, and they manifest themselves in the workplace as faulty protective features, hidden hazards, fragile assets, error traps, and misdirected values. These workplace vulnerabilities usually exist long before the unwanted consequences come to fruition, which suggests that they can be corrected proactively. It also means that events are organizational failures.

Appendix 3: Answer Key for "Checks for Understanding"

CHAPTER 1, WHAT IS A CRITICAL STEP?

1. Which of the following actions with a handgun is a CRITICAL STEP?
   a. Loading the firearm with bullets
   b. Cocking the firearm—pulling the hammer back
   c. Pointing the muzzle at a target
   d. Moving the safety lever off SAFE
   e. Pulling the trigger

   Answer: e. Pulling the trigger. Pulling the trigger is the only option that involves an irreversible and immediate transfer of energy and the movement of matter, a bullet. All other options are reversible.

2. Which attribute of a CRITICAL STEP is most important?
   a. The harm is irreversible.
   b. The harm is immediate.
   c. The harm is intolerable.

   Answer: c. The harm is intolerable. The distinguishing characteristic of a CRITICAL STEP is the severity of harm that would result from a loss of control of work. The consequence of a paper cut is immediate, irreversible, but not intolerable.

3. True or False. Is donning safety equipment, such as hardhats, eye and ear protection, gloves, a CRITICAL STEP?

   Answer: False. Though important, donning PPE provides protection, but nothing happens. There is no harm if either is done improperly, such as putting a hardhat on backwards. Later in Chapter 4, you will recognize that donning of PPE is a Risk-Important Action.

CHAPTER 2, THINKING ABOUT HUMAN PERFORMANCE RISK

1. While pulling the starter cord from the top of the engine of an old-style lawnmower, one foot is under the blade housing.
   a. Does a pathway for harm exist? If so, what is the asset and the hazard?
   b. Are there any touchpoints? If so, what are they?

   Answer: a. Yes, a pathway for harm exists for the foot (asset) under the blade housing exposing it to possible severe injury when the engine starts, and the blade (hazard) begins to rotate. b. There are two touchpoints. First, the individual places his/her foot under the housing, exposing the foot to the rotation of the blade (line of fire), and second, the person pulls the starter cord, which starts the engine and rotation of the blade.

2. True or False. A pathway for electrocution exists when an electrician is about to touch an exposed conductor with a test probe while performing a voltage measurement on an energized 120 vac circuit.

   Answer: True. When the electrician places the meter leads on the terminals of the energized circuit, there is potential for electrical shock if a finger touches an energized terminal, or the meter leads are inadequately insulated.

3. Yes or No? Walking down a long flight of stairs is a series of CRITICAL STEPS?

   Answer: Yes. The person walking is the asset (head and limbs), the pathway is the space between the person and the floor below, and the touchpoints are the person's feet while moving from one landing to the next while descending. Preferably, the person establishes another touchpoint, holding on to a handrail.

CHAPTER 3, THE WORK EXECUTION PROCESS

1. CRITICAL STEPS occur during which phase of the Work Execution Process?
   a. Preparation
   b. Execution
   c. Learning

   Answer: b. Execution. Work only happens during the execution phase, where people touch things. Things are different after work. CRITICAL STEPS change the state of assets preferably for the better, adding value. If a loss of control occurs at a CRITICAL STEP, harm likely ensues.

2. True or False. Learning occurred when the work team recognized a problem with the job just completed and verbally reported it to their supervisor.

   Answer. It's hard to say. You know learning occurs when behavior changes, but you don't know that until the same people perform that task again. There is a better chance learning is sustained if managers make appropriate changes in the system that influences the work to minimize the difference between *work-as-done* and *work-as-imagined.*

3. Which elements of RISK-BASED THINKING do frontline workers use during their preparation for a task? (There may be more than one answer.)
   a. Anticipate
   b. Monitor
   c. Respond
   d. Learn

   Answer: All elements. An effective prework discussion clarifies what is to be accomplished (the business goal to be achieved), and what to avoid (loss of control, harm, or lack of quality). *Anticipate* is a look to the future—what to expect when the work is done. This discussion helps the workers know what to *monitor* during execution of the work, such as CRITICAL STEPS and an asset's critical parameters. When they talk about worse case scenarios, they will consider how to *respond* by exploring possible contingencies and pinpointing STOP-work criteria. *Learning* occurs by recalling previous experiences with the work-at-hand and how to ensure success/avoid similar outcomes.

CHAPTER 4, RISK-IMPORTANT ACTIONS

1. What are the five conditions that must be ascertained by a healthcare worker before administering a drug to a patient?

   Answer: In the healthcare industries, the following conditions are verified by performing the "Five Rights of Safe Medication Practices." This is an **Hu** Tool for healthcare workers. The Five Rights are:
   - The right patient
   - The right drug
   - The right dose
   - The right route
   - The right time

Note: Even if you are not a healthcare professional, you may be a patient in their care, and it would help to know these yourself as a peer-check of the person administering the drug.

2. Which of the following human actions would you consider an RIA for starting an automobile engine just after an oil change?
   a. Reinstall the oil sump drain plug after draining the old oil.
   b. Add proper amount of new engine oil to the oil sump.
   c. Check the oil level using the dipstick.
   d. All the above.

   Answer: All of the above. All actions create the necessary conditions to operate the vehicle's engine without losing proper lubrication. Without the drain plug installed, any oil in the sump would leak out. Without the proper amount of oil, there will be insufficient oil to lubricate the engine during operation. Checking oil level with the dipstick verifies sufficient oil exists in the sump. All of these actions *must* occur BEFORE starting the engine, the CRITICAL STEP.

3. What RIA(s) must be done before clicking SEND on an e-mail message to an important customer to review a high-dollar contract?

   Answer: Attaching the correct document with the correct information to the message and addressing the message to the correct person. Sending business sensitive information to the wrong recipient will diminish your reputation and you may lose a customer.

4. While preparing a meal for a 2-year-old toddler, a parent answers a telephone leaving the handle of a saucepan with boiling ingredients extended over the edge of the stovetop. What's the problem?

   Answer: While distracted from the stovetop while carrying on a conversation on the telephone, the curious toddler may reach up to grab the saucepan's handle. A pathway for harm exists between the hot contents of the saucepan and the toddler's face and arms below. You can visualize in your mind the likely consequences of such a CRITICAL STEP by the toddler. The RIA is leaving the handle extending over the edge of the stovetop. The proper performance of the RIA is to turn the saucepan's handle away from the edge of the stovetop out of reach of the toddler. Notice that the RIA is reversible *before* the CRITICAL STEP occurs.

5. Before leaving on a road trip for the holidays, what actions should you take to increase the chances of an uneventful drive to your destination?

   Answer: Check proper tire pressure and tread depth, all passengers have on their seat belts, infants and toddlers are constrained properly and in the correct seat for their size, all mirrors are set for the driver, all engine fluids are at correct level, and driver's seat is properly positioned to control the vehicle.

CHAPTER 5, PERFORMING A CRITICAL STEP

1. The RU-SAFE prework discussion guidance can be used during:
   a. Work preparation by frontline workers
   b. Work planning
   c. Work execution
   d. All the above

   Answer: All the above. RU-SAFE is primarily used during prework discussions but can be applied to any work-related situation, whether in a planning or a performance phase. The checklist helps the user think through the risk and what to do about it before acting.

2. True or False. Positive control is necessary for RIAs as well as CRITICAL STEPS.

   Answer: False. Even though RIAs are important for the safe performance of a CRITICAL STEP, a mistake with an RIA has no immediate effect on the safety of assets. However, the conditions created by a mistake at an RIA can have dire consequences if not corrected before performing the CRITICAL STEP. For example, if a parachute is folded incorrectly, it can be rejected any time before its use in skydiving. It's common practice to use some form of peer-checking or independent verification for RIAs because of their importance for subsequent CRITICAL STEPS.

3. Fill in the blank. _____ thinking is necessary for the performance of CRITICAL STEPS.

   Answer: Slow. Fast thinking is unacceptable at CRITICAL STEPS. Slow thinking is intentional and engaged, whereas fast thinking tends to be automatic and mindless. Slow thinking is enabled using **Hu** Tools.

4. If there is doubt about a CRITICAL STEP:
   a. Stop and then proceed when you feel confident
   b. Stop any transfers of energy, movements of matter, or transmissions of information and get technical help
   c. With the aid of a qualified co-worker or supervisor, verify the preconditions necessary for the CRITICAL STEP, and when convinced those conditions exist, proceed.
   d. b and c.

   Answer: d. Both b. and c. Inhibit the flow of work until the facts are ascertained regarding the necessary preconditions for the safe performance of the CRITICAL STEP. Never proceed in the face of uncertainty.

5. True or False. Frequent conversations—dialogue—with co-workers improve the understanding of true state of the technical process.

   Answer: True. The human mind is notorious for making rash assumptions and making inaccurate conclusions about reality. The preponderance of counsel always improves one's understanding of reality. Reality is what you bump into when your beliefs are wrong.

CHAPTER 6, MANAGING CRITICAL STEPS

1. True or False. Frontline workers should have the freedom to diverge from approved procedures whenever they feel it is expedient for accomplishing work goals.

   Answer: False. Procedures are the most accurate and likely the safest guidance for performing technical work. They are to be followed rigorously unless adherence would place assets in jeopardy of harm. Therefore, frontline operators must possess expertise of the technology. Although most procedures are accurate, there is no such thing as a perfect procedure. Procedures are developed with assumptions in the minds of the authors and approved by managers many months, sometimes years, before use. Workers are to follow procedures mindful of landmines. When in doubt, STOP and get technical help.

2. The goal of managing CRITICAL STEPS is:
   a. Eliminate all CRITICAL STEPS
   b. Prevent human error
   c. Accomplish business goals regardless of cost
   d. Identify and control CRITICAL STEPS
   e. Improve reliability

   Answer: d. Identify and control CRITICAL STEPS. You can't eliminate all CRITICAL STEPS. No productive work would get done. It is not the aim to prevent human error. Harm can still occur even when people follow procedures or do what they think is safe. Obviously, if you cannot accomplish the business goals without harming people, product, or property, the organization will soon be out of business. Reliability is different from safety. Reliable means that a function is consistently performed according to some standard, without consideration for harm.

3. The purpose of field observation by managers is to:
   a. Assess the activity's results against expected results
   b. See firsthand the difference between *work-as-done* and *work-as-imagined*
   c. Understand how the system supports or inhibits performance
   d. Foster conversations about risk and what must go right
   e. Provide constructive feedback to workers
   f. All of the above

   Answer: e. All of the above.

4. True or False. Managing is learning. Why/why not?

   Answer: True. The management cycle, plan-do-check-adjust, requires learning to know where performance is relative to a goal. Without learning, there are no adjustments and no progress toward a goal. Managers must seek knowledge about performance in the field and with their management systems. Otherwise, safety and productivity will suffer.

5. Fill in the blank. Frontline worker adaptive capacity improves the organization's _____ to threats of harm to assets.

   Answer: Resilience. Frontline workers are the key source of resilience in the organization. Their capacity to adapt helps *create* safety as they resolve goal conflicts, close gaps between plans and reality, addressing unknown risks.

CHAPTER 7, CRITICAL STEP MAPPING

1. True or False. All CRITICAL STEPS are touchpoints, but not all touchpoints are CRITICAL STEPS.

   Answer: True. All CRITICAL STEPS involve human actions that change the state of an asset. However, most touchpoints do not satisfy the one or more attributes of a CRITICAL STEP.

2. Which of the following attributes is the deciding factor in whether a loss of control at a touchpoint is a CRITICAL STEP?

   a. Consequence is irreversible.
   b. The harm resulting from the touchpoint is immediate.
   c. The severity of harm to an asset exceeds a certain threshold.
   d. All of the above.

   Answer: c. The severity of harm to an asset exceeds a certain threshold. All other attributes may be satisfied, but if the severity of harm is minor, then a loss of control of the particular touchpoint would not be classified as a CRITICAL STEP. However, if a loss of control does result in a serious outcome, the other attributes, immediate and irreversible, must also be met to be denoted as a CRITICAL STEP. By the way, harm once realized is never reversible.

3. True or False. RIAs can be identified for an asset before the CRITICAL STEP is known.

   Answer: False. They cannot. You must know the CRITICAL STEP to know the preconditions necessary to perform it safely. Once the CRITICAL STEP's preconditions are known, then the related RIAs can be pinpointed earlier in the operation that will establish that precondition. For example, skydiving requires leaping out of an aircraft into a freefall to the earth below. To skydive safely—to skydive twice—a precondition is having a parachute secured to your body. The preceding human action to don the parachute is the RIA.

4. True or False. All RIAs are touchpoints.

   Answer: False. RIAs neither change the state of assets nor moderate the control of a hazard. That's what touchpoints do. RIAs create the conditions for touchpoints to perform those functions. Using a firearm as an example, inserting a cartridge into the chamber of a pistol has nothing to do with the discharge of the firearm. That's what the trigger does—the touchpoint and CRITICAL STEP.

CHAPTER 8, INTEGRATING AND IMPLEMENTING Critical Steps

1. True or False. Integration focuses exclusively on developing compliance with procedures, expectations, and regulatory requirements.

    Answer: False. Integration focuses on promoting a way of thinking, an operating philosophy that reinforces Risk-Based Thinking (anticipate, monitor, respond, and learn) throughout the organization. Integration depends on aligning the system to reinforce resilience and the principles of H&OP in daily operations. Emphasis is placed on mindfulness to ensure successful performance and to respond to surprise threats to assets rather than a strict compliance-oriented approach to operations.

2. Which organizational functions should consider implementing Critical Steps? (Mark all that are correct.)

    a. Operations
    b. Finance
    c. Maintenance
    d. Human resources
    e. Management/administration
    f. Supply chain
    g. Engineering
    h. All the above
    i. None of the above

    Answer: h. All of the above. All organizational functions have situations that involve Critical Steps, whether their work involves transfers of energy, movements of matter, or transmissions of information. There are some human actions, even in administrative functions, which must go right first time, involving mostly transmissions of information. Operations and maintenance are characterized as mostly hands-on, involving mostly transfers of energy and movements of matter.

3. Fill in the blank. A business case considers the _____, _____, and _____ to the organization for the initiative in question.

    Answer: Benefit (of implementing the initiative), Cost (to implement the initiative), and Risk (of not implementing the initiative)

4. A proof of concept should deploy Critical Steps in an organizational unit that is:

    a. Operations focused
    b. High-risk
    c. Generally successful
    d. All the above

    Answer: d. All the above. Critical Steps are associated with functions that are mostly hands-on in nature, such as operations and maintenance. To prove the concept's usefulness to the organization, it is important to see improvement in high-risk work activities. If an organization unit that involves low-risk activities or is a low performer, it would be difficult to detect any appreciable change in performance, even if the initiative was successful.

5. True or False. A needs-assessment of current work practices relevant to
   CRITICAL STEPS should precede the development of a change management
   playbook.

   Answer: True. A needs-assessment would reveal current strengths, weaknesses, opportunities, and threats. This information would serve as the
   starting point for the deployment of CRITICAL STEPS and the development of the playbook. You want to build on the strengths that currently
   support the implementation of CRITICAL STEPS (don't want to lose those
   features), while addressing the weaknesses that would inhibit implementation of or work at cross-purposes with CRITICAL STEPS.

Index

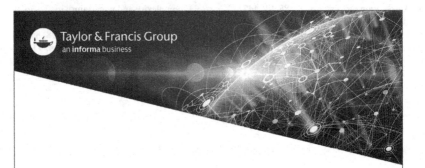

Printed in the United States
by Baker & Taylor Publisher Services